SLOTHS

William Hartston graduated in mathematics at Cambridge but never completed his PhD in number theory because he spent too much time playing chess. This did, however, lead to his winning the British Chess Championship in 1973 and 1975 and writing a number of chess books and newspaper chess columns.

When William and mathematics amicably separated, he worked for several years as an industrial psychologist specializing in the construction and interpretation of personality tests. After ten years writing a wide variety of columns for the *Independent*, he moved to the *Daily Express*, where he has been writing the Beachcomber column of surreal humour since 1998. In addition to writing about chess, he has written books on useless information, numbers, dates and bizarre academic research, including sexology.

Recently, his skills at sitting on a sofa watching television have been appreciated by viewers of the TV programme *Gogglebox*, but he has still not decided what he wants to be when he grows up.

Also by William Hartston

SLOTHS

A
Celebration
of the
World's
Most
Misunderstood
Mammal

WILLIAM HARTSTON

Atlantic Books
London

First published in hardback in Great Britain in 2018 by Atlantic Books,
an imprint of Atlantic Books Ltd.

This paperback edition first published in Great Britain in 2019
by Atlantic Books.

1 2 3 4 5 6 7 8 9

A CIP catalogue record for this book is available
from the British Library.

Paperback ISBN: 978-1-78649-425-2
E-book ISBN: 978-1-78649-424-5

Printed in Italy by 🐎 Grafica Veneta

Atlantic Books
An Imprint of Atlantic Books Ltd
Ormond House
26–27 Boswell Street
London
WC1N 3JZ
www.atlantic-books.co.uk

I'm more than half in love
with easeful sloth.

Christopher Hampton, *The Philanthropist*

CONTENTS

Note

Before we start, we have to settle one contentious matter: how should the word 'sloth' be pronounced? In his delightful book *Word Watching*, the Australian philologist Julian Burnside devotes one of his essays to the Deadly Sins, and mentions that Sloth is less frequently mentioned than the other six. 'This,' he suggests, 'might be due in part to ambivalence about its pronunciation: does it rhyme with "both" or "moth"? *OED2* rhymes it with only one, namely "both". (I would never have imagined that sentence possible.)'

Actually, the *Oxford English Dictionary* only applies that stricture to British English: it lets the Americans pronounce it either way.

I doubt that sloths are bothered about it at all.

FOREWORD

Sloths are one of the planet's most misunderstood creatures. Saddled with a name that speaks of sin and ridiculed for centuries owing to their lethargic lifestyle. For too long we humans – busy, bi-pedal apes intent on moving faster than Nature intended – have idolized animals like the cheetah. But I believe we have much to learn from the sloth and its energy saving ways, which is why I founded the Sloth Appreciation Society. With the planet in crisis thanks to our unsustainable speedy lives, it is time for us to welcome the dawn of a new era: the age of sloth. So slow down, put your feet up and read all about why the sloth is the true king of the jungle (napping between chapters is not only permitted but wholeheartedly endorsed).

Lucy Cooke
Founder, Sloth Appreciation Society
www.slothville.com

INTRODUCTION

My passion for sloths began with a YouTube video. It was one of several filmed at the Sloth Orphanage in Costa Rica, and gave me, for the first time, a close-up look at these adorable creatures. One could hardly imagine anything better designed to capture the affections. Their obvious vulnerability, apparently trusting nature, and mouth set in a permanent smile is irresistible.

Over the next few days and weeks, I ogled as many sloth videos as I could find, and my interest in these strange animals grew, but I still knew very little about them. They hung upside down, they slept a lot, and that was about it. Then, by chance, I happened to meet a zoologist. I have forgotten why I met him or what we were meant to be talking about, but I quickly came to the point and asked him what he knew about sloths. And, joy of joys, he knew a great deal.

I then explained that I was asking because I and many others had been captivated by the videos from the Costa Rica orphanage and I thought that sloths were going to be the next designer pet.

'I very much hope not,' the zoologist said.

'Why not?' I asked. 'They're adorable.'

'Baby sloths are indeed very appealing,' he said, 'but when they grow up, they can be the most vicious animals known to mankind.'

I was astonished and asked for more details.

The sloth, he told me, is very vulnerable to predators when it comes down its tree and reaches ground level. It is well designed for hanging upside down and even scampering around in rainforest canopies, but at ground level its only means of locomotion is to drag itself along with its arms. This is not an effective way to run away from predators.

Sloths have therefore evolved another means of defending themselves: they have very long, sharp claws which they can use to swipe animals that threaten them. 'I have seen,' the zoologist told me, 'a photo of the body of a jaguar which had been disembowelled by a sloth.'

Pausing a moment to sympathize with the offspring of the jaguar, which I felt must have been traumatized when their mother told them that Daddy had been killed by a sloth – such news can leave a young jaguar needing therapy for years – I rapidly changed my opinion of sloths and my admiration and respect for these creatures grew still higher. They hang upside down in the rainforest canopies of Central and South America eating the leaves around them; they only shift position when the food runs out or to come down the tree once a week to poo and pee; and they disembowel anything that irritates them. That, I felt, is the perfect lifestyle.

I resolved to find out everything I could about sloths, and the more I found out, the more I became impressed and entranced. This book is the result. Designed to satisfy the needs of sloth-lovers who want to know more about these fascinating creatures, I hope it will also recruit more people to the sloth's ever-growing band of admirers.

William Hartston
Cambridge, 2018

Chapter 1

A SLOTH BY ANY OTHER NAME

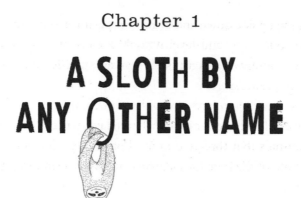

'Nothing irritates me more than chronic laziness in others. Mind you, it's only mental sloth I object to. Physical sloth can be heavenly.'

Elizabeth Hurley

I blame the Portuguese.

The word 'sloth' has been in the English language meaning 'slowness' since the twelfth century at least. Formed in much the same way as 'width', meaning 'wideness', and often spelt 'slowth' or 'sloath', it was not used to refer to a particular animal until the mid-fifteenth century when, for reasons that are very unclear, it became used as a collective noun for bears. 'Sleuth' was also used by some writers for a company of bears, but as there is no explanation for either a sleuth or sloth of bears, it is unclear which came first or whether one was an error for the other.[*]

---

[*] There is, incidentally, no generally accepted collective noun for sloths. I have seen both 'bed' and 'slumber' suggested to fill that role, but neither seems to be supported by any reputable authority.

The first reference in English to the animal we now know as sloth by that name was, according to the *Oxford English Dictionary*, in 1613 in a work by the Anglican clergyman Samuel Purchas entitled *Purchas His Pilgrimage: or Relations of the World and the Religions observed in all Ages and Places discovered, from the Creation unto this Present*. The book was intended to celebrate the diversity of God's creation and consisted of a collection of travellers' tales told to him by sailors. Since Purchas had, by his own proud admission, never travelled as much as 200 miles from Thaxted in Essex, where he was born, the tales were thus necessarily second-hand, but they referred to 'A Treatise of Brazil, written by a Portugall which had long lived there'. His description of the animals of Brazil included a thoroughly derogatory reference to something the Portuguese called *priguiça*, which Purchas translates as 'laziness':

> The Priguiça (which they call) of Brasill, is worth the seeing; it is like a shag-haire Dog, or a Land-spaniell, they are very ougly, and the face is like a woman's evill drest, his fore and hinder feet are long, hee hath great clawes and cruell, they goe with the breast on the earth, and their young fast to their bellie. Though ye strike it never so fast, it goeth so leasurely, that it hath need of a long time to get up into a tree, and so they are easily taken; their food is certaine Fig-tree leaves, and therefore they cannot bee brought to Portugall, for as soone as they want them they die presently.

Actually it is rather doubtful that Purchas was suggesting 'sloth' as the English name for the animal. The above quotation comes from the fourth (1625), hugely expanded edition of his work, which does not refer to the animal as a 'sloth' at all. That word appears, as the *OED* says, in the first (1613) edition, in which he refers to 'a deformed beast of such slow pace, that in fifteene dayes it will scarse goe a stones cast. It liueth on the leaues of trees, on which it is two dayes in climing, and as many in descending, neither shouts nor blowes forcing her to amend her pace.' Next to this, in a sidenote, he says: 'The Spaniards call it (of the contrary) the light dog. The Portugals Sloth. The Indians, Hay.'

Other later writers also give the native American words for the animal as '*aie*' or '*ai*', which is supposedly indicative of the cry of a sloth in distress (or female sloth's mating call – we shall discuss the sounds made by a sloth later). Why Purchas changed his translation of the Portuguese word for the animal from 'sloth' to 'laziness' is a mystery, but over the course of the seventeenth century, practically every European language had adopted a similar word for the animal. Purchas's book was a great influence at the time; indeed, his *Pilgrimage* was the very book that Samuel Taylor Coleridge fell asleep reading before he woke up and wrote his classic poem 'Kubla Khan'. 'In Xamdu did Cublai Can build a stately palace,' as Purchas put it, which Coleridge turned, almost 200 years later, into: 'In Xanadu did Kubla Khan / A stately pleasure-dome decree.' Of all the 4,000 pages of Purchas's great work, this influence on Coleridge must be what he is most remembered for, but his rudeness about sloths was also mimicked by others.

The first to slag off sloths in English was London clergyman Edward Topsell, who called the sloth a bear-ape or Arctopithecus in his *History of Four-Footed Beasts* (1607):

There is in America a very deformed beast which the inhabitants call *Haut* or *Hauti*, & the Frenchmen *Guenon*, as big as a great Affrican Monkey. His belly hangeth very low, his head and face like unto a childes, as may be seen by this lively picture, and being taken it wil sigh like a young childe. His skin is of an ash-colour, and hairie like a Beare: he hath but three clawes on a foot, as longe as foure fingers, and like the thornes of Privet, whereby he climbeth up into the highest trees, and for the most part liveth of the leaves of a certain tree being of an exceeding height, which the *Americans* call *Amahut*, and thereof this beast is called *Haut*. Their tayle is about three fingers long, having very little haire thereon, it hath beene often tried, that though it suffer any famine, it will not eate the fleshe of a living man, and one of them was given me by a French-man, which I kept alive six and twenty daies, and at the last it was killed by Dogges, and in that time when I had set it abroad in the open ayre, I observed, that

The sloth according to Topsell.

although it often rained, yet was that beast never wet. When it is tame it is very loving to a man, and desirous to climbe uppe to his shoulders, which those naked Amerycans cannot endure, by reason of the sharpenesse of his clawes.

Actually, Topsell got most of that information from a book called *Icones Animalium* (1552) by the Swiss naturalist Conrad Gesner, including the part about a sloth given by a Frenchman and kept alive for twenty-six days. In view of this, it seems unlikely that Topsell ever saw a sloth himself. Gesner also acknowledged that his information did not come first-hand, but he did mention, which Topsell chose to ignore, that the claws of a sloth are 'longer than those of a lion or any of the wild beasts known to us'. Both Gesner and Topsell wrote their books before Purchas, so the latter could be said to have been continuing an already established, let's-be-rude-to-sloths tradition, which exerted a strong influence throughout the sixteenth and seventeenth centuries. The real problem, however, began with Pliny the Elder in the first century AD.

Pliny's *Naturalis Historia* was perhaps the first encyclopedia of the natural world. Its thirty-seven books, divided into ten volumes, covered everything from agriculture to zoology, from painting and sculpture to mathematics and bee-keeping, setting a pattern for all other encyclopedists to follow for at least a millennium and a half. Indeed, it was one of the first ancient texts to be published in Europe, appearing in Venice in 1469, which was not long after the invention of the printing press around 1440.

Pliny, however, was not one to be inhibited by a lack of knowledge, and when he was writing of things outside his personal experience, he was liable to conflate truth with myth and observation with hearsay, which is precisely what the early writers on sloths had no compunctions about doing.

The real trouble with sloths was that they were American, while the writing of natural history had been dominated by Europeans from before Pliny to long after Columbus. As we have seen, and shall soon have confirmed by further examples, the leading European naturalists wrote a lot of rubbish about sloths, but science itself was still in its early days. Despite Pliny's *Naturalis Historia*, the term 'natural history' was not seen in the English language until 1534 (though it had been preceded by 'natural science', which included physics and chemistry, since 1425). The word 'zoology' did not arrive in the language until Robert Boyle used it in 1663, and 'scientific method' was first referred to in 1672. So perhaps we should not be too harsh on these gentlemen for being unscientific, when they did not really even have anything going by the name of 'science' to be unscientific about.

Among the earliest writers on New World animals in general and

> ...'Weariness
> Can snore upon the flint,
> when resty sloth
> Finds the down pillow
> hard'
>
> William Shakespeare, *Cymbeline* (1610)

sloths in particular were Gesner (1516–65), whom we have already mentioned, and the Dutchman Carolus Clusius (1526–1609), who was also known as Charles de l'Écluse. They both drew very fanciful pictures of sloths, which confirm that they almost certainly never saw one but were relying on the descriptions of others.

Among those descriptions may well have been the writings of the Spanish historian Gonzalo Fernández de Oviedo y Valdés (1478–1557), who spent many years in Central America, and several Portuguese missionaries and businessmen who wrote of their strange encounters in Brazil. Their various descriptions of sloths read like a game of Chinese whispers with the truth distorted in various different ways. Some seem to have concocted their stories in large part from the writings of others; some seem to have caught some fleeting glimpse of sloths themselves, but to have filled in the details with tales they have been told; while others are so vague or incorrect, it suggests their sloth experience is very limited. The French writers André Thevet (1516?–92) and Jean de Léry (1534–1613), for example, both wrote that sloths have human faces, with Thevet more precisely suggesting that it was the face of a child. Léry pointed out that sloths had never been seen eating, so he concluded that they live on air.

In 1560, the Spanish Jesuit José de Anchieta correctly pointed out that they fed on leaves. He also said they were slower than snails and had a woman's face. This last point was expanded by Fernão Cardim (1549–1625), who said it was a very ugly face 'like a badly touched woman'. Unlike the others, who all depicted sloths as walking upright on all fours like other quadrupeds, Cardim said that they walked with their belly on the ground, very slowly. Most confusingly of all, perhaps, the German explorer George Marcgrave

(1610–44) gave by far the most detailed and accurate description, complete with measurements, but he reproduced one of Clusius's drawings of the animal which made it look like a sheep with a human face.

The above descriptions can be found in a 2016 paper entitled 'Sloths of the Atlantic Forest in the Sixteenth and Seventeenth Centuries' by Danielle Moreira and Sérgio L. Mendes in the *Annals of the Brazilian Academy of Sciences*. They end the paper with a well thought-out explanation of the reasons behind their findings:

> This era of pre-Linnaean zoology was not fundamentally interested in the accurate investigation of nature itself. Instead, it followed the traditions of Renaissance classicism, which emphasizes the author's stories and knowledge. At that time, faunal records tended to focus on the symbolic meaning of the animals represented, rather than attempting to reflect precisely their zoological reality.

In the eighteenth century, when this era of symbolic nonsense-peddling should have been coming to an end, it was the French who continued the process of delivering incorrect views of sloths, but more authoritatively than ever. In 1749, the French naturalist Georges-Louis Leclerc, Comte de Buffon, wrote this character assassination:

> The sloth, which is called *ai*, or *hai*, by the natives of Brasil, on account of the plaintive cry of *ai*, which it continually sends forth, seems likewise to be confined to the new continent. It is... scarcely so quick in his motion as the turtle; it has

but three claws on each foot, its fore legs are longer than its hind ones, it has a very short tail, and no ears.

The animals of South America, which alone properly belong to this new continent, are almost all without tusks, horns, and tails; their figure is grotesque, their bodies and limbs ill proportioned, and some, as the ant-eaters, sloth, &c. are so miserably formed, that they scarcely have the faculties of moving or of eating; with pain they drag on a languishing life in the solitude of a desart, and cannot subsist in the inhabited regions, where man and powerful animals would have soon destroyed them.

In proportion as Nature is lively, active, and exalted in the ape species, she is slow, constrained, and cramped in the sloths. These animals have neither incisive nor canine teeth; their eyes are dull, and almost concealed with hair; their mouths are wide, and their lips thick and heavy; their fur is coarse, and looks like dried grass; their thighs seem almost disjointed from the haunches; their legs very short and badly shaped; they have no soles to their feet, nor toes separately moveable, but only two or three claws excessively long and crooked downwards, which move together, and are only useful to the animal in climbing. Slowness, stupidity, and even habitual pain, result from its uncouth conformation. They have no arms either to attack or defend themselves; nor are they furnished with any means of security, as they can neither scratch up the earth nor seek for safety by flight, but confined to a small spot of ground, or to the tree under which they are brought forth, they remain prisoners in the midst of an extended space, unable to move more than three feet in an hour; they climb with difficulty and pain; and their plaintive and interrupted cry they dare only utter by night. All these circumstances announce their wretchedness, and call to our mind those imperfect sketches of Nature, which, having scarcely the power to exist, only remained a short time in the world, and then were effaced from the list of beings. In fact, if it were not a desart country where the sloths exist, but had been long inhabited by man and powerful animals, they would not have descended to our time; the whole species would have been destroyed, as at some future period will certainly be the case. We have already observed, that it seems as if all that *could* be, *does* exist; and of this the sloths appear to be a striking proof. They constitute the last term of existence in the order of animals endowed with flesh and blood. One more defect and they could not have existed. To look on these unfinished creatures as equally perfect beings with others; to admit final causes for such disparities, and from thence to determine Nature to be as brilliant in these as in her most beautiful animals, is only looking at her through a straight tube, and making its confines the final limit of our judgment... the degraded species of the sloths are, perhaps, the only creatures to whom Nature has been unkind, and the only ones which present us the image of innate misery and wretchedness.

The French naturalist Georges Cuvier, who was the man responsible for the discovery that the extinct megatherium was a type of sloth, described the Bradypus in his 1817 book *The Animal Kingdom* as 'a species in which sluggishness, and all the details of the organization which produce it, are carried to the highest degree... The arms are double the length of the legs, the hair on the head, back and limbs is long, coarse and non-elastic, something like dried hay, which gives it a most hideous aspect.'

Even Samuel Johnson fell into the same pattern in his *Dictionary of the English Language* (1755), defining sloth as 'An animal of so slow a motion, that he will be three or four days at least in climbing up and coming down a tree'.

Is it possible that all these derogatory writings about the poor sloth were in some way influenced by its being named after one of the Seven Deadly Sins, or is that just a piece of English language rudeness? The evidence suggests otherwise as the following sad table of the words for sloths attests:

LANGUAGE	SLOTH (ANIMAL)	MEANING
Afrikaans	Luiaard	idler
Bulgarian	Ленивец	idler, lazy person
Croatian	Ljenivec	idler, lazy person
Danish	Dovendyr	lazy or indolent beast
Dutch	Luiaard	layabout
Finnish	Laiskiainen	lazy one
French	Paresseux	sluggish or lazy one
German	Faultier	lazy animal
Greek	Bradipous	slow foot

Hungarian	Lajhar	laziness
Icelandic	Letidyr	lazy animal
Italian	Bradipo	slow foot
Norwegian	Dowendyr	lazy or indolent beast
Polish	Leniwiec	idler, lazy person
Romanian	Lene	idleness
Russian	Ленивец	idler, lazy person
Spanish	Perezoso	lazy one
Swedish	Sengångare	slow-pace
Turkish	Tembel hayvan	lazy beast

With the exception of Italian and Greek, which have adopted versions of the scientific 'Bradypus' (meaning 'slow-foot') and the Swedish 'slow-pace', every one of those words for 'sloths' has connotations of laziness, idleness or the Deadly Sin of Sloth. Even the exotic-sounding Turkish just means 'lazy animal'.

There can hardly be an animal on earth with such an unfortunate, derogatory name. From aardvark to zebra, from baboon to yak, the vast majority of creatures have been given English names that are exclusive to them. The name of an aardvark may come from the South African Dutch for 'earth pig', 'orang-utan' may be the Malay for 'person of the forest', 'rhinoceros' may derive from the Ancient Greek for 'nose-horn', an axolotl may take its name from the Aztec for 'water' and 'dog', yet in all such cases the animal's name has been assimilated into English (and any other languages) to refer to itself alone.

Recognizably descriptive animal names are rare by comparison. Anteaters may eat ants, a blackbird may be a bird that is black (except the females, which are brown), hummingbirds may hum, woodpeckers peck wood, grasshoppers hop through the grass, yet such descriptive terms are rare, particularly among mammals, and to find anything with a name as uncomplimentary as that of a sloth, we have to descend to the level of bedbugs and slugs.

So indolence and laziness are reflected in an immediately identifiable manner in the very name given to sloths almost everywhere around the world. Yet slow movement need not always be associated with laziness. One of the slowest yet least slothful people I met during my ill-spent youth as a chessplayer was a Dane named Sloth. His first name is Jørn, of which the pronunciation is closer to 'yearn' than 'yawn', but is still perhaps the most appropriate first name for a Sloth. He is also, as far as I know, the only Sloth to win a world championship.

When I met him, Jørn Sloth was one of the most promising young chessplayers in Denmark but his main passion was not the usual form of chess but Correspondence Chess, in which players send their moves to opponents by post. This is perhaps the slowest game on earth and demands a phenomenal level of precision and patience. Games may last for years and often slow to a rate of one move a month.

In 1980, Jørn Sloth, at the age of thirty-five, became the youngest player ever to win the World Correspondence Chess Championship. The final of the competition had begun in 1975, but when I met him in 1978, Sloth told me he was on the verge of winning it. He had a winning position in a game against his nearest rival, but it would take him another eighteen moves to finish the Russian off. At the current rate of postal services between Denmark and the USSR, he expected it to last about another year and a half.

I was well acquainted at the time of the misery of trying to defend a hopeless position for an hour or two before finally forced to resign, but the notion of such misery extending for a year and a half was enough to put me off the idea of correspondence chess entirely.

Sloth, incidentally, was the 174th most common surname in Denmark in 2004 but only the 149,328th most common in the 2000 US Census. I do not know why there are so many Sloths in Denmark, but the Sloth I knew was as fine an example as one could wish to find of the potential benefits of apparent slothfulness.

Besides being a World Champion, Jørn Sloth, though he probably does not know it, was also responsible for one of the best excuses I ever heard for losing a game of chess. That excuse, I should emphasize, was not made by Sloth himself, who was far too rational and modest a man to make excuses, but was offered by an English player who had just been defeated by him. 'He kept putting me off by hanging upside down from the board,' he said.

Long before the achievements of Jørn Sloth, however, there was one writer who came close to capturing the wonder of sloths. He was Charles Waterton (1782–1865) , the Squire of Walton Hall, who was as fine an example of benevolent English eccentricity as one could wish for. His *Wanderings in South America* (1825) contained an observant first-hand account of the behaviour of sloths which may have been the first time anyone had anything nice to say about them:

His looks, his gestures and his cries all conspire to entreat you to take pity on him. These are the only weapons of defence which Nature hath given him. While other animals assemble in herds, or in pairs range through these boundless wilds, the sloth is solitary and almost stationary; he cannot escape from you. It is said his piteous moans make the tiger relent and turn out of the way. Do not then level your gun at him or pierce him with a poisoned arrow – he has never hurt one living creature. A few leaves, and those of the commonest and coarsest kind, are all he asks for his support...

Those who have written on this singular animal have remarked that he is in a perpetual state of pain, that he is proverbially slow in his movements, that he is a prisoner in space, and that, as soon as he has consumed all the leaves of the tree upon which he had mounted, he rolls himself up in the form of a ball and then falls to the ground. This is not the case. If the naturalists who have written the history of the sloth had gone into the wilds in order to examine his haunts and economy, they would not have drawn the foregoing conclusions. They would have learned that, though all other quadrupeds may be described while resting upon the ground, the sloth is an exception to this rule, and that his history must be written while he is in the tree.

The sloth, in its wild state, spends its whole life in trees, and never leaves them but through force or by accident. An all-ruling Providence has ordered man to tread on the surface of the earth, the eagle to soar in the expanse of the skies, and the monkey and squirrel to inhabit the trees: still these may change their relative situations without feeling much inconvenience; but the sloth is doomed to spend his whole life in the trees, and, what is more extraordinary, not *upon* the branches, like the squirrel and the monkey, but *under* them. He moves suspended from the branch, he rests suspended from it, and he sleeps suspended from it. To enable him to do this he must have a very different formation from that of any other known quadruped... Hence his seemingly bungled conformation is at once accounted for; and in lieu of the sloth leading a painful life, and entailing a melancholy and miserable existence on its progeny, it is but fair to surmise that it just enjoys life as much as any other animal, and that its extraordinary formation and singular habits are but further proofs to engage us to admire the wonderful works of Omnipotence.

With those words, the eccentric Charles Waterton brilliantly put his finger on precisely what all earlier writers had missed: the sloth is unlike other creatures. It is indeed disastrously designed to live the way we do, striding upright on the ground and moving as quickly as possible from place to place, but has evolved for a truly alternative lifestyle. And having been around for more than 50 million years, compared with about 7.5 million for genus Homo and at most 300,000 for *Homo sapiens*, it has had plenty of extra time to become better adapted to its lifestyle than we have to ours.

Chapter 2

TWO TOES OR THREE?

'I have more in common with a three-toed sloth
or a one-eyed pterodactyl or a Kalamata olive
than I have with Winston Churchill.'

Boris Johnson, on his book *The Churchill Factor* (2014)

When you meet a sloth, the first question you should ask is this: How many toes does it have?

There are two types of sloth in the world: two-toed and three-toed. Actually that simple statement needs clarifying. First, sloth purists will insist the difference is in their fingers not toes: all sloths have three toes on their hind limbs; it's the front limbs that differ in the number of digits. I shall stick to calling them two-toed or three-toed. That alliterates better than using the word 'fingered', and in any case calling them 'two-fingered' may evoke a feeling that they are making a rude gesture at us. I could, I suppose, use the scientific terms, 'Bradypus' for the three-toed sloth and 'Choloepus' for the two-toed, but Bradypus comes from the Greek for 'slow-foot' while Choloepus is even ruder, meaning 'lame-foot'.

In German, incidentally, the two types of sloth are called *Zweifingerfaultier* and *Dreifingerfaultier* ('two-finger' and 'three-finger lazy animal'), but English speakers have always been reticent about discriminating between an animal's front and hind limbs by calling the extremities 'hands' and 'feet', so we are equally reluctant to talk about 'fingers' on front limbs.

Also, there are not just two types of sloth: there are two species of two-toed sloth and four species of three-toed, but it's the number of toes that makes the greatest difference, so let's start by listing the similarities: they look similar; they both live at the top of trees in the rainforest canopy; they both hang upside down from the branches; they both poo and pee only about once a week. So here's a table of some of their differences:

TWO-TOED SLOTHS	THREE-TOED SLOTHS
are 58–70 cm long and weigh between 4 and 8 kg	are about 45 cm long and weigh 3.5–4.5 kg
have long, fine, tidy hair	have scruffier, rougher hair
have big eyes	have smaller eyes
spend a good part of their lives upside down	spend around 10 per cent of their lives upside down
have only a vestigial tail	have short tails 6–7 cm long
are predominantly nocturnal	rest or sleep at night
will hiss, bare teeth and try to slash with claws when disturbed	are very docile
usually have six cervical vertebrae (neck bones)	usually have nine cervical vertebrae
either drop poo through trees or come down to poo and leave it uncovered	come down from their tree to poo, then cover it with leaves

eat buds, leaves, flowers, fruit, twig tips and in captivity even meat	are very fussy about their food, only eat leaves mostly from the Cecropia tree
often climb down trees head first	generally back down trees

We should also mention that while both types of sloth are renowned for moving slowly, the two-toed variety are a little faster than their three-toed cousins. Estimates for a sloth's speed when crawling on the ground vary between about two and four metres per minute. Its speed when climbing a tree is at the upper end of these estimates, but it can move faster when scampering upside down along a tree's branches. In all cases, the two-toed sloth is up to 50 per cent faster than the three-toed.

Perhaps most interesting of all, the two types of sloth are scarcely related to each other, having gone their separate ways down the evolutionary tree an estimated 30–35 million years ago. With that length of time to evolve, what is remarkable is how similar they are when you consider that they have lived as separate creatures for as long as dogs and cats. This is particularly striking in view of the fact that for most of that time, there were other sloths around that evolved in a completely different way, but before we return to their evolution, we should become acquainted with the six sloth species currently around.

All sloths are members of the superorder of placental mammals known as Xenarthra. The *Xen-* that starts the word is the same as the *xen-* in 'xenophobia', fear of foreigners. It comes from a Greek word meaning 'strange' or 'foreign'. The *-arthra* is the same as in 'arthritis', and refers to joints. So Xenarthra in general and sloths in particular are named for their strange joints. In the sloth's case, this refers to the joints in the hip (though it might just as well refer to the bones of the neck). The curious, reinforced way in which the sloth's legs are connected to its hips are what gives it the flexibility to cling to a branch with its legs and swivel its body around at unbelievably gymnastic angles. Within Xenarthra, which also includes armadillos and anteaters, all sloths are members of the order Pilosa, which just means 'hairy', and the suborder Folivora, which means 'leaf-eater'.

Curiously, all the xenarthrans used to be included, along with pangolins and aardvarks, in an order called Edentata, which literally means 'toothless'. In the case of sloths, however, it is only the front teeth they are lacking, which makes it very confusing to classify them as Edentata. The pangolins and aardvarks are therefore nowadays given their own order while sloths are simply called Xenarthra.

We are now down to a level in the taxonomy that contains only sloths, of which there are six species. Two-toed sloths are divided into two species: Linné's sloth and Hoffmann's sloth. Three-toed sloths may be the brown-throated sloth, pale-throated sloth, maned sloth or pygmy sloth.

Linné's sloth is named after the great Swedish botanist, zoologist and physician to the Swedish royal family Carolus Linnaeus (1707–78), who in 1761 was ennobled as Carl von Linné. Three years earlier, in 1758, he had given the name *Choloepus didactylus* to a species of sloth found in Guyana and the Amazonian forests of Brazil, Colombia, Peru, Ecuador and Venezuela.

The other species of two-toed sloth is named after the German physician and naturalist Karl Hoffmann (1823–59), who was born in Stettin (now Szczecin in Poland) and studied at Berlin University.

In 1853, he travelled to Costa Rica to collect plant and animal specimens and liked it so much that he settled in its capital, San José, where he opened a medical clinic and pharmacy which also sold wine and liquor to boost his finances. As well as giving his name to *Choloepus hoffmanni*, he also identified species of woodpecker, parakeet, millipede and a species of ant-eating thrush, all of which now bear his name. Hoffmann died of typhus in Puntarenas, Costa Rica at the age of thirty-five. His name was first attached to the sloth species by Wilhelm Peters, curator of the Berlin Zoological Museum, in 1858, which was exactly a hundred years after Linnaeus had given a taxonomic classification to the other two-toed sloth species.

Hoffmann's sloth lives in Bolivia, Brazil, Colombia, Costa Rica, Ecuador, Honduras, Nicaragua, Panama, Peru and Venezuela, and looks very similar to Linnaeus's species, though DNA analysis suggests that the two species diverged six or seven million years ago. The main ways to differentiate between them by sight are their slightly different head shapes and the fact that *Choloepus hoffmanni* seems to spit more.

Having named these two species after noted naturalists, the zoological world seemed to decide that they would do better naming three-toed sloths after physical characteristics. So these are separated, as we have seen, into the brown-throated, pale-throated, maned and pygmy varieties.

The brown-throated sloth, which is also known as the Bolivian sloth, may be identified, you will not be surprised to hear, from the dark patch of thick fur over its throat. Their other name, however, is a little misleading, as they can be found in many other countries of Central and South America as well as Bolivia.

Pale-throated sloths do not have uniformly pale throats, but their name comes from a pale yellow spot in that region. It is also easier to tell the difference between male and female in this species of sloth than in others, for males have a yellow-orange patch on their backs with a black stripe going down its centre. More endangered than the brown-throated or pale-throated species, the maned sloth used to be widespread but is now generally seen only in the wet forests of south-eastern Brazil. The mane that gives it its name is a wide circle of long, dark hair around its neck. Their eyes are also surrounded by dark fur.

Last, smallest, and by far the least numerous of all, the pygmy sloth is only found on one island off the coast of Panama. We shall have more to say about them later, in Chapter 12, on conservation.

Returning to the sloth's evolution, in the world of hunter versus prey, there are three basic ways to survive: you can grow bigger than your predators so that you can fight them off; you can move faster than them so that you can run away when threatened; or you can develop effective ways of hiding from them. Our conclusions, based on observation of present-day sloths and fossil records of earlier sloths, provide good case histories of the first and third of these. With the hindsight of evolutionary history, we can say that for sloths, the run-away-and-hide strategy has proved to be the most successful, despite the fact that both bulk and speed offer advantages both to predator and prey.

Until around 10,000 years ago, the sloth world was dominated by giant ground sloths which rampaged mainly around Central and South America. The fossil evidence does not yet provide a complete picture of sloth evolution, but the evidence suggests that sloths first emerged at least 50 million years ago and grew bigger and bigger.

According to research published in 2014, the largest of the eighty identified species of sloth, the megatherium, grew to the size of an elephant, six metres tall and weighing up to seven tonnes. Its claws alone could be a foot long. Over its long period of evolution, the megatherium increased in weight, putting on an average 129 kg per million years.

Needless to say, these massive creatures did not hang upside down from the topmost branches of the rainforest canopy, but at the same time as they were galumphing around the plains, or at least for the last 12 million years or so, other smaller sloth species evolved. With remarkable foresight (or more likely good evolutionary luck), their predilection was for hiding rather than brawn. Rather than fighting or running away from predators, the ancestors of the sloths we know today climbed slowly up trees and hung there, moving only when absolutely necessary. Their hiding skills even led to a remarkable form of camouflage, attained by letting algae grow on their fur. This gave them a greenish appearance that blended in beautifully with the foliage around them, making them hard to detect, these days, even by eagles looking for a quick slothy snack.

To judge by the fossil evidence, however, the small arboreal sloths were a bit of an eccentricity while their huge cousins were around. Small sloth skeletons from that period are hugely outnumbered by the enormous ones, though this may be mainly due to the fact that fossils of arboreal creatures are less likely to be preserved than those of land-dwelling creatures. Indeed, so few fossils of small tree-dwelling sloths have been found that opinions differ on the precise evolutionary relationship of the arboreal sloths to the ground sloths. While all agree that all sloths, be they two-toed, three-toed or giant ground sloths, had a common ancestor, some say the two-toed sloth evolved from giant sloths, and others maintain that they evolved independently.

In his *Origin of Species* (1859), Darwin wrote this in pre-emptive reply to those who might question his radically new ideas:

> It may be asked in ridicule whether I suppose that the megatherium and other allied huge monsters, which formerly lived in South America, have left behind them the sloth, armadillo, and anteater, as their degenerate descendants. This cannot for an instant be admitted. These huge animals have become wholly extinct, and have left no progeny. But in the caves of Brazil there are many extinct species which are closely allied in size and in all other characters to the species still living in South America; and some of these fossils may have been the actual progenitors of the living species.

The evidence, from both fossils and DNA investigation, indicates that Darwin may have been more or less correct.

Although some recent discoveries suggest that the ground sloths may have survived in a few isolated areas until around 4000 BC, it is clear that they, along with most of the other megafauna, mostly disappeared in the late Pleistocene era some 5,000 years earlier. Within a relatively short period, all sorts of giant creatures became extinct, including mammoths, giant beavers, giant tortoises, lions, mastodons, sabre-toothed salmon and giant sloths. The reasons for this are unclear but the prime suspects are either climate change or over-hunting by humans. Another possibility is that the extinction was caused by some long-term effect of continental drift. After all, sloths both small and large had existed for a very long time before the formation of the Panama isthmus joined North and South America around three million years ago. We have very little idea what effect this had on animal life or what diseases may have then been able to move between the Americas.

Whatever caused it, over 90 per cent of sloth species were wiped out, leaving only the small arboreal sloths. It was almost as if their cautious strategy of hiding at the tops of trees was finally justified by enabling them to hide even from a continent-wide extinction event.

Remarkably, the postal services of at least five countries have issued postage stamps featuring giant sloths. Appropriately, the first seems to have been Cuba in 1982, which is thought by some to be the last place in which they became extinct. In 1958, they produced a stamp depicting the skeleton of a giant sloth, and in 1982 an artist's impression of the animal itself. Cuba was followed by Guyana (1991) and, most surprisingly, Cambodia (1994), which was a long way from anywhere sloths or giant sloths have ever been.

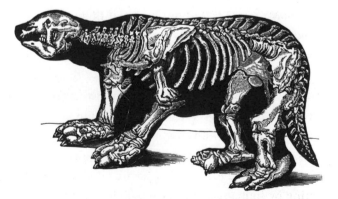

Illustration of a giant sloth skeleton from *Meyers Konversations-Lexikon*, 1897.

I should mention one further delightful account of the reason for the disappearance of the giant sloth. Mark Twain, in his *Letters from the Earth* (written in 1909 but not published until 1962), devoted Letter Five to an account of Noah's preparations for the Great Flood. After pointing out the vast number of species of animals and insects Noah had to gather together on the ark, he tells us how Noah did as well as he could, but, just as he was preparing to cast off and set to sea, he received a disturbing report of vast numbers of creatures heading his way: 'Sloths as big as an elephant; frogs as big as a cow; a megatherium and his harem huge beyond belief; saurians and saurians and saurians, group after group, family after family, species after species – a hundred feet long, thirty feet high, and twice as quarrelsome.' And to make matters worse, they were all very hungry, 'and would eat up everything there was, including the menagerie and the family'. So Noah set off without them.

'All these facts were suppressed,' Mark Twain tells us, 'in the Biblical account. You will find not a hint of them there. The whole thing is hushed up.' And that is why there are no giant sloths, or dinosaurs, around today.

Before we leave the ground sloth to its undisturbed giant extinction, there are three further questions we need to answer:

- What makes us believe these huge creatures were related to sloths?
- What does a US President have to do with it?
- And how do avocados fit into the picture?

The man responsible for the first of those was the French naturalist and zoologist Georges Cuvier (1769–1832), whose early research into findings of ancient bones led to his being called the 'father of palaeontology'. He was the first person to establish extinction as a genuine phenomenon rather than just a speculative theory, and it was his description in 1796 of a megatherium skeleton that first identified similarities between its bones and teeth and those of sloths, anteaters and armadillos.

Indeed (and this is where the US President fits in), when some prehistoric bones were found in a cave in Virginia in the mid-1790s they were sent to Thomas Jefferson for possible identification. Quite apart from being one of the Founding Fathers, the main author of the Declaration of Independence, and a future Vice-President and then President of the United States, Jefferson was a noted palaeontologist. It is difficult to imagine anyone whose opinion on the matter was more likely to be highly esteemed. On 10 February 1797, Jefferson published his 'Memoir on the Megalonyx' for the American Philosophical Society, listing all the bones he had received, his measurements, and the details of other bones found at the site of which he had been notified. He listed the bones sent to him as follows:

1. a small fragment of the femur or thigh bone;
2. a Radius, perfect;
3. an Ulna or fore-arm, perfect, except that it is broke in two;
4. three claws, and half a dozen other bones of the foot, but whether of a fore or hinder foot is not evident.

'These bones,' he wrote, 'only enable us to class the animal with the Unguiculated [having one or more claws] Quadrupeds, and, of these, the lion being nearest to him in size, we will compare him with that animal.' After some calculations on the relative sizes of the bones he was examining and those of a lion, and assessments of the likely relative weights of the animals, he went on: 'I will venture to refer to him by the name Great-claw, or Megalonyx, to which he seems sufficiently entitled by the distinguished size of that member.'

After writing this 'Memoir' and preparing to send it off, Jefferson received a periodical from London which caused him to add a PS:

> After the preceding communication was ready to be delivered in to the society, in a periodical publication from London I met with an account and drawing of the skeleton of an animal dug up near the river la Plata in Paraguay [which] receives the new denomination of Megatherium... The Megatherium is not of the cat form, as are the Lion, tyger and panther, but is said to have striking relations in all part of its body with the Sloth (Bradypus), tatoo (Dasypus), pangolin (Manis) and the anteaters (Myrmecophaga and Orycteropus).

The periodical he had received was the September 1796 issue of *Monthly Magazine* which included a description of a megatherium by Georges Cuvier, whom we have already met and of whom we shall have more to say later.

It was not until 1822 that the scientific community got round to accepting a name for the bones Jefferson had examined, when, at

> *'Desidiae valedixi; syrenis istius cantibus surdam posthac aurem obversurus'* – I bid farewell to Sloth, being resolved henceforth not to listen to her siren strains.
>
> Samuel Johnson, Diary (October 1729)

the suggestion of the French zoologist Anselme Gaëtan Desmarest (1784–1838), it was called *Megatherium jeffersonii* – 'Jefferson's megatherium'. Later it was discovered that it was different from the megatherium giant sloths, so was renamed *Megalonyx jeffersonii*, so Thomas Jefferson is still the only US President to have had a giant sloth named after him. The mystery about giant sloths and avocados, however, is another matter entirely.

Together with the bones and claws of ground sloths, researchers discovered in many cases samples of their poop or even stomach contents. These revealed that they were vegetarian, with a particular liking for avocados. In fact the era of the megafauna was a golden age for these large-stoned fruits. Unlike most creatures today, giant sloths had large enough mouths and throats to swallow avocados whole. They would then digest the fleshy part of the fruit and excrete the stone.

This was very profitable for the avocado as a dispersal mechanism for its seed. Instead of dropping to the ground, where all the seeds from one tree would have to compete for the same resources, they now took root over a large area and avocado trees grew everywhere. Then the giant sloths became extinct. It is now known that avocados have been part of the Mexican diet for some 10,000 years, but those were wild plants. The earliest evidence of human cultivation of avocados dates back at most 5,000 years when the Inca, Olmec and Mayan civilizations of Central America all grew avocados. It was the Aztecs, incidentally, who gave the fruit the name *ahuacatl*, meaning 'testicle', which its shape supposedly reminded them of.

That cultivation takes us back to around 3000 BC, which was several thousand years after the extinction of ground sloths. There have been several erudite academic discussions of such 'seed dispersal anachronisms', but it is still a bit of a mystery how the avocado survived the extinction of the giant sloth.

The latest mystery surrounding giant sloths, however, is far bigger than an avocado.

In 2017, a paper was published in the journal *Ichnos* regarding the discovery of large numbers of huge burrows in sedimentary rock in South America. Such burrows had been known since the 1930s or longer but had been dismissed as of unknown origin, but were thought probably to have been caused by some sort of geological process, or possibly were tunnels dug by early humans, which may have begun as caves, and had been developed to serve as short-cuts or protective hideaways.

With no evidence as to which of these theories was correct, the burrows were largely ignored or forgotten until the last ten years, when a discovery of a massive example in the state of Rondonia in north-west Brazil aroused great interest. When geologists investigated it in 2010, they found it was twice the size of any previously discovered cave or burrow, and since then over 1,500 such structures have been found. The largest of them are up to two metres high, four metres wide, and extend for a length of up to a hundred metres.

Two things led to a drastic reassessment of these burrows. First, their smoothness and apparently purposeful design increasingly ruled out their having been produced by any natural process. In any case, nobody could suggest a natural process that might have been responsible. Second, there were scratch marks on the walls and roof of the tunnels that suggested they had been made by an animal's claw.

The size and shape of the marks were found to match the claws of giant armadillos, in the case of smaller tunnels, and giant sloths in the case of larger ones, and the height and width of the tunnels also matched those creatures. So were the most massive underground structures dug by giant sloths? And if so, for what purpose? Could it have been to escape predators, or some aspect of climate change?

On the plus side for the sloths-as-architects/construction-workers theory, it has been pointed out that around half of all current mammalian species live a semi-fossorial existence, dividing their time between burrows, for resting, and the outdoors for eating. Between 3 and 4 per cent live their entire lives underground. In view of this, it seems at least plausible that their ancestors also spent a good deal of time in burrows. Another recent suggestion is that the primary purpose of the burrows was to offer a safe refuge from predatory smilodons, a type of sabre-toothed tiger, which were around at the same time.

On the minus side, however, insufficient animal bones, fossils or other signs of giant-sloth occupation have been found in these burrows to be sure they were the architects. But perhaps they were too house-proud to leave such slowly biodegradable messes in their homes.

In May 2018, a DNA analysis of bones found in a giant sloth burrow in Chile went a long way towards resolving one question of sloth evolution. By comparing modern sloth DNA with that of *Mylodon darwinii* (the giant sloth discovered by Charles Darwin), the researchers concluded that today's two-toed sloths are evolutionary sisters of Darwin's discovery, the two species diverging about 22 million years ago, while all modern three-toed species are at best distant cousins of *Mylodon*, splitting away even earlier.

Chapter 3

THE SLOTH THAT SAVED DUBLIN

And Other Tales of Nineteenth-Century Sloth Rehabilitation

> **World Peace could be at hand – if only we could be more like the sloth.**
>
> William N. Thompson, *Parables from (a not quite) Paradise* (2003)

You could say that it was a sloth that saved Dublin Zoo in 1844. It is unlikely that you would find anyone to agree with you on that point, but you may still say it, for it is undeniable that a sloth did at least play a part in saving Dublin Zoo from financial disaster at that time. Here are the facts.

Zoos, or 'menageries' as they were usually called, had existed since ancient times. In Egypt, Queen Hatshepsut is said to have established a collection of African animals around 1500 BC, while Emperor Wen Wang, founder of the Zhou dynasty of China, displayed his power and wealth by keeping a collection of animals from all parts of his realm in the eleventh century BC. For centuries, however, such collections were designed to be viewed only by royalty or the imperial court.

In Europe also, zoos were not designed to be viewed by the masses. The Menagerie at the Tower of London was very much a royal affair, beginning in 1204 when King John ordered three shiploads of wild beasts to be brought from Normandy after the English king had ceded control of the territory to France. It was enhanced by gifts of three leopards from Holy Roman Emperor Frederick II to Henry III in 1235, a polar bear from King Haakon IV of Norway in 1251, and an elephant from Louis IX of France in 1255. This menagerie could be viewed by visitors to the Tower of London during the reign of Queen Elizabeth I but the spread of zoos was discouraged by a seventeenth-century edict restricting the privilege of exhibiting exotic animals to the Keeper of His Majesty's Lions at the Tower of London.

Despite this, a number of travelling animal shows grew up in Britain, of which the best known was Gilbert Pidcock's menagerie, which found a home at the Exeter Exchange in the Strand in London, where it opened to the public in 1793. Pidcock died in 1810 and control of the menagerie passed to his deputy, Edward Cross, who was both an experienced animal keeper and a great showman. Cross became owner of the menagerie in 1817 and opened it from 9 a.m. to 9 p.m. daily, when the public could see three main exhibits for a shilling, or fork out two shillings to view everything. The peak time for visitors was 8 p.m., when the animals were fed and the elephant, named Chunee, rang a loud bell to signal the beginning of feeding time.

Chunee met a sad end in 1826 after going berserk on his daily walk down the Strand and killing a keeper. He had previously been a docile animal, even appearing on stage in plays and pantomimes at the Theatre Royal Covent Garden with the celebrated actor Edmund Kean. Towards the end of his life, however, he had become increasingly violent, possibly because of toothache caused by a rotten tusk. After the disaster in the Strand, soldiers were summoned from Somerset House to put Chunee down. Even after being hit by 152 musket balls, he was still alive and had to be finished off with a sword (or possibly harpoon) wielded by a keeper.

Even though hundreds paid their shilling entrance fee to the Exeter Exchange to see Chunee's body being dissected by doctors from the Royal College of Surgeons, this had a bad effect on subsequent visitor numbers to Cross's menagerie which closed down in 1828. Even in its best years, however, there had been complaints about disruption in the Strand caused by animal noises and the awful smell coming from the building that housed the animals. Nevertheless, its overall effect was to increase the desire of Londoners, and many wider afield, to see elephants, lions, giraffes and other exotic beasts of which they had only read. The 1830s and 1840s thus became a period of great activity in planning and opening zoos in Europe, and that brings us back to the Dublin sloth.

It is unclear whether Cross's menagerie included a sloth. Some accounts say it did, but it seems more likely this was a sloth bear from India rather than a true sloth from the Americas. We know that some of the animals from the Exeter Exchange were sold to the London Zoological Society, while others were moved to Cross's own Surrey Zoological Gardens, but there is no evidence these included a sloth, which comes into the story in Dublin in 1844.

Dublin Zoo had opened its doors in 1831 with forty-six mammals and seventy-two birds, all donations from the London Zoological Society. By that time, the Zoological Gardens in Regent's Park had already been open for three years, but only as a collection for scientific study. It would not be opened to the public until 1847. Berlin Zoo was also not opened until 1844 and while Vienna Zoo at Schönbrunn Palace, which had begun life as an Imperial private collection in 1752, had allowed the general public to visit since 1779, entry was permitted only to those who were 'dressed properly'.

By the mid-1840s, Ireland was facing increasing economic problems that culminated in the Potato Famine. With around half the population already subsisting mainly on potatoes, the potato blight and the Great Hunger it led to in Ireland gave people far more serious problems to think about than a trip to the zoo, and the finances of the Dublin Zoological Society were looking dire.

In 1837, however, Dr Robert Ball had been appointed honorary secretary of the Society and he proved to be just the man for the job. His fondness for sloths was recalled by his son, Sir Robert Stawell Ball, who went on to become Astronomer Royal:

> A sloth, on one occasion, arrived in the evening, and in order to reproduce the climate of a Brazilian forest as nearly as was possible at such short notice, the sloth was hung on the back of a chair before a fire in the dining-room. I have no doubt the animal passed a very comfortable night. I well recollect how long afterwards we used to point out the marks of its claws and teeth on the back of the chair... Years later, the

Two types of sloth imaginatively depicted in J. G. Stedman's *Voyage to Suriname* (1799)

A Milodon giant sloth statue at Puerto Natales, Chile

Argentinia celebrated the giant
sloth in a 2001 postage stamp

Giant sloth exhibit at the Museum
of Zoology in Cambridge

A giant sloth and an iguanodon hold the Cambridge University coat of arms

Skeleton of a three-toed (fingered) sloth at the Museum of Zoology in Cambridge

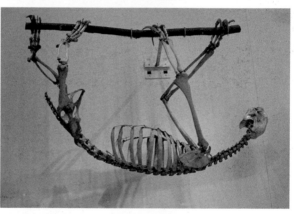

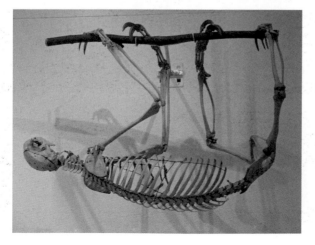

Skeleton of a two-fingered sloth

Perfectly designed for hanging upside down

Choloepas claws on fingers and feet: 2+3=5

Sloths can swim twice as fast as they walk

Sloths' legs are not designed for standing upright

Upwardly mobile

Remarkable head rotation saves energy

Just hanging around

Young sloth confused at crossroads

Hoffman's two-toed sloth: blonde hair and long snout

chair, which still bore the creature's teeth and claw marks, made a very interesting after dinner story. As I have since heard, my father held the sloth to be rather a delicate animal. Consequently, it was sent as a present to the Zoological Gardens in Regent's Park, whereupon the council of that Society sent us a young giraffe.

That account was only published in 1915 in a collection of *Reminiscences and Letters of Sir Robert [Stawell] Ball* by his son, W. Valentine Ball, grandson of the sloth-loving Robert Ball. More contemporary accounts of the giraffe-for-sloth transaction indicated that the giraffe was obtained by Dublin in exchange for 'a two-toed sloth' and a 'tiger-cat' (a small wildcat, probably some sort of ocelot) from Central America. Further details were given in the annual report of the Dublin Zoological Society, as published in *Freeman's Journal and Daily Commercial Advertiser (Dublin)* on 7 May 1845:

The most remarkable addition to the collection this year has been the Giraffe, 'Albert', presented to the society under the following circumstances:– One of your secretaries (Mr Ball) being in London in May last, represented to influen- tial members of the Zoological Society there how greatly a grant of a giraffe would be valued in Ireland, not only by your society, but by the public at large in Dublin; and he stated how much beyond the means of the society the purchase of so costly an animal was. The council of the London Society met, and in the most prompt, handsome and unanimous manner made a free gift of the very beautiful giraffe now in your possession. They selected this animal with the kindest consideration as the most portly and hardiest of their stock, although from its being the first reared in Europe, being so far acclimatized and a personal favourite and pet of many of the society, he was the most valued, and probably most valuable, giraffe in London. Your council could return no equivalent for this princely donation, but with marked thanks on the part of the society they had the gratification of being able to add two animals to the collection in the Regent's Park, not before seen alive in London – a two-toed sloth (*Choloepus unau*) and a new species of tiger-cat, named by Mr Ball *Felis melanuva*. The extremely rapid growth of the giraffe (about thirty inches in the year), and the probability of his growing five or six feet more, makes it imperatively necessary to build a house suited for him. The temporary one which he at present occupies, being only 14 feet high, his horns suffer from striking against the ceiling.

The 'Occurrences at the Garden' daily record of happenings at Regent's Park Zoo offer a few more details of the transaction. The entry for 23 May 1844 reports the death of a great-eared owl, the illnesses of two lionesses and a parakeet and: 'Rec'd, a sloth lent by the Zoological Society of Dublin; an ocelot presented by the Zoological Society of Dublin'. A few weeks later, on 14 June 1844, they record: 'Giraffe born 27 May 1841 presented to Dublin'. The gift of the ocelot and loan of the sloth is confirmed in the Minutes of Council of the Royal Zoological Society of Ireland for 5 June, which refer to 'a specimen of the two-toed Sloth' being sent to the London Society's Menagerie 'for a limitless period'.

Whatever the details of the exchange, the giraffe was just what Dublin Zoo needed. Encouraged by an inspired decision by Robert Ball to lower individual entry prices from sixpence to a penny (but only on Sundays after 2 p.m.), large numbers of people flocked to see Albert the giraffe, and the zoo's financial problems were considerably eased. Their Council Minutes for 3 July record that it was 'resolved that the very marked thanks... be given to the Council of the Zoological Society of London for their most munificent donation of the beautiful giraffe named Albert'. They also voted unanimously 'that the two-toed Sloth be presented to the Zoological Society of London, conceiving that the present would be acceptable, and likely to be serviceable to science'. Some may say that Dublin Zoo was saved by a giraffe, but I like to feel that the sloth played a vital part too.

Ball was not so lucky with another sloth mentioned in his son's *Reminiscences*, which includes the tale of a less fortunate member of the species bought by his father. In 1847, we are told, Sir Robert Ball bought a sloth for £15. 'I am afraid it is a bad bargain,' he writes, 'as he has a cold and is sick already.' It is unclear whether this was the same animal as the sloth that hung over the back of a dining chair in front of the fire.

Meanwhile, back in London, the sloth was proudly welcomed by both the zoo and its visitors. On 27 May 1844, the Personal Column on the front page of *The Times* newspaper carried the following announcement from the Zoological Society of London:

> The Fellows of the Zoological Society are
> informed that a SLOTH *bradypus tridactylus*,
> the first ever imported into England, has been

received at the gardens of the Society, in the
Regent's park, and will be exhibited after today.

The announcement was issued in the name of barrister and
naturalist William Ogilby (1808–73), who was the secretary of
ZSL. On 28 May, the *London Standard* reported initial reactions to
the new sloth: 'The juvenile portion of the visitors were delighted
with the antics of the ourang outang, whilst the more aged were
taken with the slow motions of a sloth (bradypus tridactylus), the
first ever imported into England.' Clearly, the writers of both these
reports, however, had not had the opportunity to count the number
of toes on the animal's front legs: it was *didactylus*, not *tridactylus*.

Not only was the first half of the nineteenth century a good
time for zoos in general, but as a side effect it was a good time for
the public understanding of sloths in particular. As we have seen,
writings before that time tended to be based on a mixture of hearsay
and misunderstanding, leading to the sort of nonsense produced by
the Comte de Buffon and Baron Georges Cuvier. As zoos began
to give more people the chance to observe sloths for themselves,
even if not in their natural habitat, the general feeling towards these
animals became more positive.

In his 1830 book *The Menageries*, published by The Library of
Entertaining Knowledge, the Scottish naturalist James Rennie
(1786–1867) wrote about sloths with remarkable prescience:

The sloth is usually described as slow in his movements, and
as in a perpetual state of pain; and from his supposed inaction
his name is derived. And why is this? He had not been seen
in his native woods by those who described him: he was

resting upon the floor of some place of confinement. His feet are not formed for walking on the ground; they cannot act in a perpendicular direction; and his sharp and long claws are curved. He can only move on the ground by pulling himself along by some inequalities on the surface, and therefore on a smooth floor he is perfectly wretched. He is intended to pass his life in trees; he does not move or rest upon the branches, but under them; he is constantly suspended by his fore-legs, and he thus travels from branch to branch, eating his way, and sleeping when he is satisfied. To put such a creature in a den is to torture him, and to give false notions of his habits. If the sloth be placed in a menagerie, he should have a tree for his abode; and then we should find that he is neither habitually indolent nor constantly suffering.

Curiously, Rennie was exactly half right. His argument was predominantly against menageries, because they remove animals from their natural settings and enclose them instead in very limited environments, which can give completely the wrong impression to observers. While he was right about the apparent 'constant suffering' of sloths being caused either by the observers or by the limitations of the circumstances under which sloths were being observed, his comment about 'habitual indolence', if somewhat derogatory, was rather misguided. Sloths in the wild really are remarkably inactive most of the time, particularly the three-toed varieties. Two-toed sloths match them for inactivity only during the day; at night they wander around and even move from tree to tree much more.

Despite the reasonable accuracy of descriptions such as Rennie's, however, artists still tended to use rather too much of their own imaginations when portraying sloths, as the cover of an 1835 issue of the *Penny Magazine* shows:

'The Sloth': front cover of *Penny Magazine*, 3 January 1835.

Meanwhile, back in Dublin, I do not know how long the sloth had been exhibited to the public, and it may even have always resided for much of the time with the Ball household. When it reached

London, however, it certainly aroused interest, and articles began to be published giving accounts of this animal, never before seen in Britain, based on direct observation that at last contradicted the misinformed and prejudiced words of the French zoologists.

On 29 June 1844, only a few weeks after the animal arrived in England, the *Illustrated London News* published an article under the title 'The Sloth at the Zoological Gardens',* beneath a rather fetching woodcut of the animal hanging upside down from a branch. 'The collection of the Zoological Society in the Regent's Park has lately been enriched by the acquisition of a new wonder of the animal creation,' it began, going on to say that it was 'a living specimen of the three-toed sloth (Bradypus tridactylus) the first that has ever reached this country alive'.

As the illustration made very clear, it was in fact a two-toed sloth, but apart from that mistake, the article was very perceptive, particularly when it came to pointing out the 'lack of personal observation' that had bedevilled earlier accounts of sloths,

> among others, of Buffon's celebrated, eloquent but altogether fallacious dissertation in which he bewails the fate of the poor sloth as a bungled and misshapen monster quite unfitted for any enjoyment of life and destined from the first moment of his existence to continual misery and pain. But more careful observers have shown that this apparently incongruous animal is as beautifully adapted for the sphere Nature intended it to occupy as any other of her myriads of creatures.

---

* The Zoological Gardens had opened in 1828, and the abbreviation 'zoo' was first seen in 1835. In popular usage, it caught on quickly, but in formal writing, it was still 'the Zoological Gardens' for some time.

The unnamed writer of that piece went on to describe how the anatomy of the sloth is perfectly designed for hanging upside down and passing effortlessly from one tree to another without touching the ground, even though its forelegs and hind legs are 'formed and jointed to the body as utterly to incapacitate them from supporting it on earth'.

Not long before that article was published in the *Illustrated London News*, the members of the Zoological Society of London had themselves been briefed on their new arrival. As the *Proceedings of the ZSL* for 28 May 1844 reported:

> The Secretary called the attention of the Meeting to a specimen of the Two-toed Sloth, Bradypus didactylus, which was now in the Gardens, and requested Mr. Ball, Secretary to the Royal Zoological Society of Ireland, to communicate such particulars connected with the habits and manners of this curious animal as had fallen under his observation.
>
> Mr. Ball regretted that it was out of his power to state the exact locality from which the animal had been obtained; however, he had reason to believe that it was brought from Demerara [in British Guiana]. Its general food was sea-biscuit and water; of fruit it partook sparingly, but he had observed it pick the young buds of the hawthorn flowers and eat them with great avidity.
>
> While in the Zoological Gardens at Dublin its favourite position was where it was supported partly by the branch to which it clung, and partly by an adjoining branch on which its back could rest.
>
> In lapping water, the great length to which its tongue was

protruded was very remarkable, thereby showing its affinity to the other Edentata of South America.

In July 1844, the London sloth appeared for the first time in the *List of the Animals in the Garden of the Zoological Society of London*, where it appeared as 'The Two-Toed Sloth Bradypus didactylus, Linn'. 'This is the first individual of the species ever imported to England,' the Society proudly added. 'It is supposed to have been procured from Demerara and was presented to the Society by the Royal Zoological Society of Ireland. In fine weather, this animal is allowed to range on the large trees outside of the building.' These were probably the trees around the giraffe house. The picture shown by the *Illustrated London News* in celebration of the sloth's arrival must have been the most lifelike the British had ever seen.

The London Zoo sloth as depicted in the *Illustrated London News* in 1844.

Almost a year later, in May 1845, the wonders of the London Zoo sloth were told, in similar tones, to readers of the *Penny Magazine*, published by the Society for the Diffusion of Useful Knowledge, under the heading 'Notes Upon a Living Sloth in the Gardens of the Zoological Society'. But this time, though, it was accompanied by a far more accurate picture of a sloth than the one they had displayed so proudly a decade earlier. 'For a considerable period,' the piece began, 'a living sloth has existed in the Gardens of the Zoological Society; and as this is the first instance, so far as is known, of this singular creature being brought alive to this country, a few remarks upon the animal, drawn from actual observation, may not be uninteresting.'

After that modest start, however, the author laid into the opinions of earlier writers, bolstering his case with allusions to Shakespeare's *Richard III* ('sent into this breathing world...') and *Hamlet* ('Nature's journeymen'):

> It is but within the last few years that the misrepresentations respecting [the sloth] have been cleared away, – misrepresentations in which writers on zoology, with little philosophical acumen, have followed each other, taking their key-note from the florid but erroneous details of Buffon. A moment's reflection, indeed, on the acknowledged fact that these animals, natives of the vast and dense forests of South America, are essentially arboreal in their habits, might have induced naturalists to pause before they set them down as cripples 'sent into this breathing world but half made up;' and *that* lamely and clumsily, as if 'Nature's journeymen' had made them and not made them well. Such ideas are inconsistent

with philosophy; and, besides, not a little presumptuous, inasmuch as they involve the admission that the Creator may fail in the workmanship of his hands... So far, however, is the sloth from being the wretchedly deformed creature which it has been represented, that it is one among thousands of examples which might be particularly selected as an instance of design and wisdom.

To prove his point, the writer went on to explain the manner in which the sloth 'displays the utmost ease travelling from branch to branch, suspended by its hooks (for into such are its claws modified), with singular address; and when climbing, throwing itself into various singular attitudes, indicative of perfect security'.

Yay! Go, sloths!

For the most supremely elegant put-down of Buffon's description of sloths, however, we had to wait until 1925, when William Beebe, a great writer on sloths himself, brought out an anthology of the best writings on natural history under the title *The Book of Naturalists*. He included in this a tale from Charles Waterton's 1825 *Wanderings in South America* concerning a meeting with a sloth. In his introduction to the tale, Beebe wrote as follows:

> The selection on sloths herewith presented is a just refutation of the opinion of the great savant Buffon, who, in spite of his genius, fell into the most grievous error in his estimation of a sloth. He says: 'The inertia of this animal is not so much due to laziness as wretchedness; it is the consequence of its faulty structure. Inactivity, stupidity, and even habitual suffering

result from its strange and ill-constructed formation. Having no weapons for attack or defence, no mode of refuge even by burrowing, its only safety is in flight... Everything about it shows its wretchedness and proclaims it to be one of those defective monsters, those imperfect sketches, which Nature has sometimes formed, and which, having scarcely the faculty of existence, could only continue for a short time and have since been removed from the catalogue of living beings. They are the last possible term amongst creatures of flesh and blood, and any further defect would have made their existence impossible.'

If we imagine the dignified French savant himself naked, and dangling from a lofty jungle branch in the full heat of the tropic sun, without water and with the prospect of nothing but coarse leaves for breakfast, dinner, and all future meals, an impartial onlooker who was ignorant of man's normal haunts and life could very truthfully apply to the unhappy scientist, Buffon's own comments. All of his terms of opprobrium would come home to roost with him. A bridge out of place would be an absolutely inexplicable thing, as would a sloth in Paris, or a Buffon in the trees.

Quite apart from its noble achievement in rehabilitating sloths in the eyes of Europe, the London Zoo sloth's move from Dublin in 1844 was extremely fortuitous as the Potato Famine was on the verge of reducing Ireland to a desperate state. In Albert the giraffe's first year in Dublin, the zoo received 144,000 visitors, which was equal to about half the population of Dublin. The income from these visitors left the zoo in a decent financial position, but in the years from 1845 to 1850, when the Potato Famine was at its worst, Dublin Council had great problems feeding the animals. Following a ruling that they should not be given any food that could be eaten by people, the zoo was reduced to using horse-meat and rough barley in place of bread and oats. Some greens were still bought for the ostrich but Albert the giraffe had to supplement his diet with onions, clover and rock salt. Gastronomically speaking, the sloth did well to move to London.

The sloth, incidentally, which apparently remained unnamed, lived for four years, seven months and twenty-three days in Regent's Park, before its death was solemnly announced in the zoo records as having taken place on 16 January 1849. By that time, it had done its job. In 1847, an encyclopedia publication entitled *Wonders of the Animal Kingdom* was produced 'under the direction of the Committee of General Literature and Education, appointed by the Society for Promoting Christian Knowledge, London'. Its entry on 'Sloth' was over 3,000 words long and began with the following character reference for this once maligned creature:

> The sloth, unlike any other quadruped, makes its home in the trees, never leaving them except from force or accident. It does not rest upon the branches like the squirrel and the monkey, but clings to the under part of them. Many naturalists, not having studied the habits of this singular animal in its native condition, rashly thought, from the melancholy cry it uttered, and from its inability to move without great difficulty on level ground, that its life was one of constant pain. It received its present ill-deserved name from an opinion,

that the animal was too indolent to stir from the tree it had chosen for its abode, until it had eaten all the leaves, and then, compelled by hunger, it rolled itself into the form of a ball and dropped to the ground, from whence with slow and awkward movements it regained another tree to strip in the like manner.

Later and more observant naturalists have shown how erroneous these opinions are.

As I said before – Go, sloths!

Chapter 4

NEW SLOTHS FOR OLD

In Brazil, looking up at the sloths in repose, I felt
I was in the presence of upside-down yogis deep in
meditation or hermits deep in prayer, wise beings
whose intense imaginative lives were beyond the
reach of my scientific probing.

Yann Martel, *Life of Pi* (2001)

Around the same time as one sloth was taking great steps for sloth-kind by flitting from Dublin to London, an intense debate was raging over the place of sloths in world history. I might have said 'the place of sloths in evolution', but this was more than a decade before Darwin published his *Origin of Species* in 1859 and, even then, the word 'evolution' did not occur at all in the first edition (though 'evolved' appears once as the very last word of the book).

Before Darwin, zoology was as much a matter of theology as science. When Buffon (1707–88) wrote his defamatory nonsense about sloths, it was part of his 44-volume *Natural History*, of which thirty-six volumes were published in his lifetime and another eight after he died. At the time, it was generally believed by educated people that all the creatures on earth were exactly as God had ordained at the Creation. Buffon took the view that it was absurd to think that an all-powerful God could be bothered 'with the way a beetle's wings should fold'. His belief was that God indeed created the many different species for their various purposes, but as they went forth and multiplied, they changed as time went on.

His opinion of sloths fitted beautifully into his general 'Theory of American Degeneracy', arguing that animals in America were generally much more feeble than their European counterparts. 'No American animal can be compared,' he wrote, 'with the elephant, the rhinoceros, the hippopotamus, the dromedary, the camelopard [giraffe], the buffalo, the lion, the tiger, etc.' His views about the native people of the Americas were even worse, and by today's standards appallingly racist:

> In the savage, the organs of generation are small and feeble. He has no hair, no beard, no ardour for the female. Though nimbler than the European, because more accustomed to running, his strength is not so great. His sensations are less acute; and yet he is more timid and cowardly. He has no vivacity, no activity of mind.

Naturally, the Americans did not take this lying down. Amusingly, Thomas Jefferson had the skin and antlers of a moose sent to Buffon, as well as the antlers of deer, caribou and elk, and the skin of a panther, just to show Buffon that New World quadrupeds were just as good as those in the Old World.

Back in Europe, reading Buffon's words, others found it difficult to explain how a perfect God had managed to create such a supposedly imperfect creature as a sloth, unless He was just showing off by demonstrating what He could get away with. Buffon got around it by simply blaming the heat and humidity for bringing about the sloth's alleged degeneracy. On practically all matters other than sloths, however, Buffon was one of the greatest and most respected scientific minds of his age and presented ideas that were a century ahead of their time.

Quite apart from his early days as a mathematician, in which he was one of the originators of the binomial theorem and its applications to probability theory, he went on to become the first person to conceive of a geological history of the earth divided into a series of epochs. The Church doctrine at the time saw the world as less than six thousand years old, but Buffon saw the seven days of Creation as seven periods of indeterminate length. His first period involved the sun being hit by a comet which created the earth and the other planets; the second period saw the earth cooling and becoming a solid body; during the third period, it was entirely covered by an ocean; then the waters subsided, volcanoes erupted, dry land was exposed, and plants and animals appeared. Finally, the original land-mass broke up to form the separate continents; and only after all that did human beings appear.

This was so strikingly at odds with the ecclesiastical view that Buffon was investigated by the Theological Faculty of the Sorbonne in Paris. He avoided censure by publishing a recantation saying not that he was wrong, but that he had not intended his description of the earth's formation to be taken as a contradiction of spiritual truths.

Perhaps his only rival at the time in taking on the task of describing and classifying all of animal life was the Swedish botanist, physician and zoologist Carl Linnaeus (1707–78) whose systematic taxonomy covering all life-forms formed the basis for the scientific nomenclature that is still used today. However, Buffon had nothing but criticism for Linnaeus's methods of classification, which he considered totally misguided. By basing his taxonomy on degrees of similarity in a variety of physical features, Linnaeus was, according to Buffon, not describing the true order of nature but imposing an arbitrary order defined by the human mind. 'This manner of thinking has made us imagine an infinity of false relationships between natural beings,' he declared. 'It is to impose on the reality of the Creator's works, the abstractions of our mind.'

> As to ourselves, we all know the speed produced by the employment of steam; we have experienced it either on railroads, or in boats when crossing the sea; but such a flight is like the travelling of a sloth in comparison with the velocity with which light moves.
>
> Hans Christian Andersen,
> 'The Shoes of Fortune' (1838)

Coming from the man who imposed the abstractions of his own mind and apparent lack of imagination on the supposed physical attributes of the sloth, this may seem a bit rich, but it was all part of Buffon's general attack on Linnaeus's system. Specifically, he wrote that 'The more one increases the number of divisions in natural things, the closer one will approach the truth, since there actually exist in nature only individuals, and the Genera, Orders and Classes exist only in our imagination.' Clearly, Buffon saw Linnaeus as little more than a train-spotter or stamp-collector, ticking items off and placing them in his albums where he thought fit. Buffon's own version was based on an early variation of evolution, in which the environment acted directly on organisms through what he called 'organic particles'. Defective environments thus led to defective creatures such as the sloth, or Americans.

Linnaeus's response to this criticism was terse and to the point. He dismissed Buffon's *Histoire Naturelle* as a work 'in French, without pretty figures, with wordy descriptions... few observations, beautiful ornate French... without any method'. Writing of the author, he snidely said that Buffon 'isn't particularly learned, but as he is rather eloquent, that seems to count for something'. He also named a genus of weed 'Buffonia', which some have taken as a specific insult to Buffon, though others say the name is from the Latin for toad, *bufo*, and that Linnaeus was spelling the genus name with a double-f before he had even heard of Buffon. No doubt, in view of his own low opinions of sloths, Buffon would have thought it entirely appropriate that the name of Linnaeus was attached to a species of *Choloepus didactylus*.

Nowadays, Linnaeus is revered as the father of taxonomy, while Buffon is almost forgotten. His rapid fall from fame, however, may be largely due to his death, at the age of eighty, in 1788, one year before the French Revolution. As a French count, and the Head of the Jardin du Roi in Paris, he would have been a prime candidate for the guillotine. His tomb, in a chapel next to the church of Sainte-Urse Montbard in eastern France, was broken into during the Revolution and the lead that covered the coffin was stolen to produce bullets. Buffon's only son, Georges, was sent to the guillotine on 10 July 1794, only eighteen days before the execution of Robespierre himself.

Buffon's baton as chief sloth persecutor was taken up by another French nobleman, Jean Léopold Nicolas Frédéric, Baron Cuvier (1769–1832), known to his friends as Georges. While happy to continue Buffon's denigration of living sloths, however, Cuvier was also responsible for earning them a new level of respect, for he was the man who established the connection between the sloths that Buffon had been so rude about, and the magnificent, lumbering ground sloths that had rampaged through the Americas thousands of years before.

Cuvier's *Le Règne Animal* ('The Animal Kingdom') was first published in four volumes in 1817, but grew to a third edition of twenty-two volumes, completed after the author's death, effectively replacing Buffon's *Natural History* as the definitive work on life on earth. Like Buffon, he rejected both evolution (or at least Lamarck's pre-Darwinian version of it, involving the inheritance of acquired characteristics) and the idea that all animals are created perfect and stay that way. He also shared with Buffon a contempt for sloths.

We have already given one example (see above, p. 12–13) of the views Cuvier expressed about sloths in his animal kingdom. Here are two more:

They have a short face. Their name originates from their excessive slowness, the consequence of a construction truly heteroclite, in which nature seems to have amused herself by producing something imperfect and grotesque.

The hind feet are obliquely articulated on the leg, and rest only upon their outer edge; the phalanges of the toes are articulated by a close ginglymus [a hinge-like joint], and the first, at a certain age, becomes soldered to the bones of the metacarpus or metatarsus, which also, in time, for want of use, experience the same fate. To this inconvenience in the organization of the extremities is added another, not less great, in their proportions. The pelvis is so large, and their thighs so much inclined to the sides, that they cannot approximate their knees. Their gait is the necessary effect of such a disproportioned structure. They live in trees, and never remove from the one they are on until they have stripped it of every leaf, so painful to them is the requisite exertion to reach another. It is even asserted that to avoid the trouble of a regular descent, they let themselves fall from a branch.

Cuvier's writings on the sloth read like a headmaster's report on a child he particularly dislikes, refusing to see any possible good in the poor creature. Unlike Buffon, however, he respected Linnaeus's classification system and employed it in his analysis of bone fragments and other fossils of extinct animals. Indeed, it was Cuvier who was the first to establish that extinction of species really occurred, and the first creatures he positively identified as extinct were the mastodon (an extinct elephant species) and the megatherium, or ground-dwelling giant sloth.

The power of Cuvier's innovative methods in comparative anatomy is illustrated by a famous story about an attempted trick played on him by his students late one night. One of the mischievous students, it is said, dressed up in a devil's costume and woke up Cuvier with the chant 'Cuvier, Cuvier, I have come to eat you!' In response, Cuvier is said to have opened his eyes to examine his visitor, after which he said, 'All creatures with horns and hooves are herbivores. You can't eat me', and went back to sleep.

Cuvier had only been twenty-six years old when he gave his first public lecture at the National Institute of Science and the Arts in Paris, and it was something of a sensation. By close examination of their bones, he demonstrated that African and Asian elephants were different species, but the bones from an elephant-like creature which had recently been unearthed in Ohio, to which he later gave the name 'mastodon', were from neither of those, and was now extinct. That name, incidentally, means 'breast teeth' and is formed of the two parts *mast-* (as in 'mastectomy') and *-odon* (as in 'odontology'), and was inspired by a similarity Cuvier perceived between the ridges on the teeth he examined and nipples.

Not long after identifying the mastodon bones, he examined the skull and teeth of other beasts dug up in America and confidently identified them as having come from early relatives of the sloth family. These bones, Cuvier announced, 'seem to me to prove the existence of a world previous to ours, destroyed by some sort of catastrophe', going on to propose a new theory of 'catastrophism' to account for the history of life on our planet. Just as Buffon had envisaged seven epochs, Cuvier thought the development of animal life passed through an unspecified number of periods, each separated from its immediate predecessor by a catastrophe destroying almost all the life that had preceded it.

Thanks to the giant sloth (and the mastodon, and the even older prehistoric fossils that were now beginning to be dug up), the Church began to take an interest again, seeing something in science they could embrace. The six days of Creation and alleged 6,000-year age of the world may still have needed a bit of rationalizing, but Cuvier's catastrophes were seen by some as proof of one account in the Bible at least: for isn't that exactly what Noah's Flood is all about?

Interestingly, Charles Darwin himself seemed to embrace that idea. In 1832, he wrote a letter to his sister Caroline Sarah Darwin, telling of his joy at finding fossil bones in Patagonia during his voyage on the *Beagle*:

> The chief source of pleasure has been to me, during these two months, from Nat: History. – I have been wonderfully lucky, with fossil bones. – some of the animals must have been of great dimensions: I am sure that many of them must be quite new; this is always pleasant but with the antediluvian animals it is doubly so.

'Antediluvian', of course, literally means 'before the flood'. And so, having God and Noah on their side, sloths began to grow a following among clerical scientists, notably the Reverend William Buckland, Doctor of Divinity, Fellow of the Royal Society, and Professor of Geology and Mineralogy at Oxford. Buckland was the first person to describe the excavation of the fossil bones of what he called 'Megalosaurus' ('great lizard'), which was the first known dinosaur. He was also renowned for a unique passion for eating any animal, and once announced that the only thing tasting worse than a mole was a bluebottle. Another tale about Buckland is that once, when lost in fog on the outskirts of London while riding in a coach, he dismounted, scooped up a handful of earth, tasted it and declared: 'Uxbridge!'

Now that his formidable credentials have been established, you will be impressed, I am sure, to hear that in 1833, William Buckland delivered a lecture to the Linnean Society of London entitled 'On the Adaptation of the Structure of the Sloths to their Peculiar Mode of Life'. He began with a statement that left no doubt as to where he stood:

> There are I believe no animals whose structure has been so generally misunderstood by naturalists, and respecting which so many errors have popular acceptance, as the Sloths. They are often quoted, even by the authorities in comparative anatomy, as affording examples of imperfect organization, and are proverbially misrepresented as holding the lowest place in creation and as constructed only to lead a life of inconvenience and misery.

He then goes on to discuss Cuvier's argument that sloth skeletons show how badly adapted they are to reality, with arm and forearm almost double the length of thigh and leg, 'so that when the animal goes on all fours he is obliged to drag himself upon his elbows, and if he attempted to stand erect upon his hind feet, the entire fore foot would still rest upon the ground' (so the sloth can never therefore 'stand upright because his hind feet are so ill articulated for walking that they are unable to support the body in such a position'), before pointing out that:

> the learned author seems to view the structure of this animal as Buffon had done before him in relation only to its defects, as ill-adapted to the ordinary movement of quadrupeds in walking upon the ground. Had he considered its peculiarities in relation to their perfections with reference to the habit of the animal living constantly upon trees and coming to the ground only for the purpose of passing from one tree to another, in those rare cases where it cannot pass from tree to tree without descending, the consideration of this habit would at once have explained all the apparent incongruities of structure, and every organ which appears so anomalous and ill adapted for walking upon the ground would have been found pre-eminently fitted to supply the wants and comforts of an animal destined to spend its life upon trees. The extraordinary length of the arm and forearm so inconvenient for moving on the earth are of essential and obvious utility to a creature whose body is of too great weight to allow it to crawl to the extremity of the branches to collect the extreme buds and youngest leaves which form its food, these long arms in fact perform the office of the instrument called lazy tongs whereby the creature brings food to the mouth from a distant point without any movement of the trunk.

The nature of the shoulder and elbow joints, the breadth of the pelvis and outward position of the thigh bones, the distance of the knees from one another, and the curvature of the leg bones were also explained by Buckland as perfectly designed for a 'quadruped which was to feed, to sleep, and in short to dwell entirely upon trees'. Systematically working through those peculiar anatomical features that Cuvier had adduced as evidence of the sloth's deliberate design to make it as difficult as possible to walk like normal animals on the ground, he showed instead how well designed they are for the business of a sloth's real life. He then quotes at length from Charles Waterton's accounts of keeping a sloth, which fully support everything he has said, before providing a telling finish:

> Does it not follow from the above comparisons of the habits of the Sloth with its form and structure that, so far from being in any respect an imperfectly constructed animal, it is fitted with admirable perfection of mechanism to its unusual habits and peculiar condition of life? It is true that if rapid locomotion be an essential attribute of a quadruped the Sloth will labour under the imputation of debility, but we have seen that agility and activity have been superfluous to an animal that has no occasion to run or walk and that the slow and torpid movements of its arms and claws cause no inconvenience to a creature whose food is stationary upon trees... The charge of imperfection therefore can with no more justice be advanced against the construction of the Sloth because its locomotive powers upon the ground are slow than against the structure of fishes because they are not furnished with legs.

Finally, there is just one more person who deserves a mention in this tale of nineteenth-century sloth rehabilitation, and that is none other than the man who invented the word 'dinosaur'. Sir Richard

Owen (1804–92) was the English biologist who founded London's Natural History Museum in 1881 and became its first director. His reputation as an anatomist earned him the right to be the first to examine the corpse of any animal that died at London Zoo (which led to a tale that his wife returned home one day to find a dead rhinoceros waiting for her).

As well as being an unsurpassed anatomist, he also had a reputation for being stubborn, intensely ambitious, jealous of the success of rivals (particularly the success of Charles Darwin), and not averse to stealing the work of others. He did, however, play an important role in sloth history.

When Cuvier, in 1796, reached the conclusion that the extinct megatherium was related to the modern sloth, Owen had not yet been born, but it was not until Owen published his exhaustive *Description of the Skeleton of an Extinct Gigantic Sloth* in 1842 that the scientific world unanimously agreed with Cuvier's previously controversial viewpoint. Owen's extraordinary work consisted of 167 pages of well-argued text covering every bone and joint of fossil giant sloths and living species of tree sloths, plus another hundred pages of tables, measurements and plates. Bone by bone, he went through the old and new skeletons, comparing them not only with each other, but with different creatures. The result finally silenced the last critics of Cuvier's findings.

Owen did, however, end with a summary that, if not complimentary to Cuvier, at least made up for it by lifting the reputation of sloths:

The genera Bradypus and Choloepus have been regarded by all zoologists as forming one of the most anomalous and isolated groups in the mammiferous class, of which no other proof is needed than the fact, that whilst Cuvier, in the *Règne Animal*, has placed the Sloths in the lowest order of Unguiculata [animals having claws or nails], his successor [Henri de Blainville] in the celebrated French school of zoology has seen reason for raising them to the highest or quadrumanous [having all four feet modified as hands] order, agreeably with an old opinion of Linnaeus.

Lastly, on the topic of sloth rehabilitation, I should mention *Observations on the Anatomy of the Sloth* (1832) in which the US naturalist Richard Harlan corrected many of the misconceptions of Cuvier. His seven-page account is written in lucid and elegant English, except for one paragraph that begins: 'The reproductive organs of this animal are singularly anomalous' and then continues in Latin. The classically educated, however, will then learn that the female sloth possesses a clitoris.

Chapter 5

ARE SLOTHS SLOTHFUL?

Slothfulness casteth into a deep sleep;
and an idle soul shall suffer hunger.

Proverbs 19:15 (King James Version)

Despite the perceptive and insightful words of Charles Waterton, it took the sloth around 250 years to recover fully from the character assassination of the Comte de Buffon and start to be taken seriously. It was only in the twenty-first century that some zoologists and naturalists began to suggest that sloths might not be slothful after all.

The trouble with Buffon's buffoonery was that he had never seen a sloth in the wild and lacked the imagination to realize that they were different from other mammals. He knew that sloths were terrible at making progress at ground level and he could see from drawings, their skeletons and other people's accounts that their feet were hopelessly designed for running, but he still assumed that they tried to walk on all fours like other quadrupeds. He also knew that they were covered with green slime and didn't seem bothered by it, so he drew the conclusion that they were filthy and unhygienic.

Waterton got it right, but his own reputation as a wealthy English eccentric may have got in the way of his being taken seriously. Quite apart from his claim to have been a direct descendant of at least eight Catholic saints, this Yorkshire squire was said to like dressing as a scarecrow and sitting in trees, impersonating his own butler and tickling his guests with a coal brush, and pretending to be a dog and biting the legs of guests when they arrived at his estate. He was also a noted taxidermist whose works included a tableau of reptiles dressed as famous English Protestants which he called 'The English Reformation Zoologically Demonstrated', and an item he called 'The Nondescript' consisting of the bottom of a howler monkey turned into the likeness of a human face.

So it was the anti-sloth views of serious French authorities such as Buffon and Cuvier which were accepted as official doctrine rather than the pro-sloth comments of Waterton. The animal, after all, is called 'sloth' so it stands to reason that it must be a lazy ne'er-do-well. Observation of sloths that had been captured and were displayed in zoos also seemed to confirm that picture. In general, they either died quickly, apparently confirming a hopeless inability to adapt, or survived but slept for at least sixteen hours a day. That's almost as lazy as a koala, and at least those Australian marsupials have the excuse of being stoned out of their tiny minds on eucalyptus leaves.

Oddly enough, while investigating aspects of the sleeping cycle of pygmy sloths on the island of Escudo off the coast of Panama in 2010, sloth researcher Bryson Voirin discovered a similarity between the brainwave patterns of this species and those of humans addicted to Valium. His investigations led to a theory that the red mangrove leaves which form their main diet may contain a fungus with Valium-like properties that drug the sloths and are the reason these pygmy sloths are so laid back, even for sloths. That possible link between sleepy koalas and sleepy sloths, however, came when we were already beginning to have a deeper respect for the reposeful nature of sloths.

The date on which attitudes towards sloths began to change may be placed precisely at 23 August 2008 when the journal *Biology Letters* published a paper entitled 'Sleeping Outside the Box: Electroencephalographic Measures of Sleep in Sloths Inhabiting a Rainforest'. The lead author of the paper was Dr Niels Rattenborg, head of the Rattenborg Research Group at the Max Planck Institute for Ornithology at Seewiesen in Upper Bavaria. In 2016, Rattenborg produced the first evidence to show that birds sleep while flying and also suggested how they might do so without running the risk of bumping into each other.

The great leap forward for sloth-kind, however, may be attributed to the second named writer of the report in *Biology Letters*, the same young researcher named Bryson Voirin who was to propose the stoned-sloths hypothesis. And his research all began with a passion for climbing trees. It was this passion that led Voirin's biological studies to specialize in arboreal animals, which quickly led him to a fascination with sloths. 'Almost everything about them is specially geared towards their arboreal lifestyle,' he said in an interview in 2009. 'They have green algae-rigged fur that disguises them as foliage. Their nerves are evolved to react slower, so they do not flinch or react to noises. These animals are true specialists.'

That 'green algae-rigged fur' came into its own in a research paper published in the online science journal *PLOS One* in 2014. Written by a team headed by Sarah Higginbotham of the Smithsonian Tropical Research Institute in Panama, its title was 'Sloth Hair as a

> # The constant Attendant on Sloth is *Sluttishness:* She who gives her Mind to Idleness, can neither be thoroughly clean in her own Person nor the House.
>
> Eliza Fowler, *A Present for a Servant Maid. Or, the Sure Means of Gaining Love and Esteem* (1743)

Novel Source of Fungi with Potent Anti-Parasitic, Anti-Cancer and Anti-Bacterial Bioactivity', and it charted some remarkable medicinal properties of sloth fur fungus.

The potential effectiveness of fungi in general as a source of antibiotics had previously been detected, but sloth fur had never itself been examined in that respect. However, as they pointed out, sloths 'carry a wide variety of micro- and macro-organisms on their coarse outer hair'. They tested fifty fungal extracts from the surface fur and found that twenty were active against at least one strain of bacteria. 'We found a broad range of activities against strains of the parasites that cause malaria (*Plasmodium falciparum*) and Chagas disease (*Trypanosoma cruzi*), and against the human breast cancer cell line MCF-7.' And the first acknowledgement at the end of the paper mentioned our favourite tree-climbing sloth-lover: 'We thank Bryson Voirin for collection of sloth hair samples.'

Before 2008, almost all the academic research on sloths had been conducted on animals in captivity. What was needed was someone who could get closer to their arboreal hiding places, but despite his expertise in shinning up trees, Voirin found it far from easy at first to catch sloths. 'They are incredibly difficult to even find,' he explained. 'I think it took three weeks to find and catch my first one. After that I got better and improvised techniques to catch them... They can move quick when they want to, and they are not defenceless'. He has scars from their claws all over his arms to prove that last point. When he did catch them, however, he did nothing more than attach something like a small bowler hat to their heads and this contained the electroencephalograms that measured their brain activity, enabling him and his colleagues to conduct the first study of how sloths sleep in the wild.

The results were astonishing. As Rattenborg's paper reported:

> We performed the first electroencephalogram (EEG) record-ings of sleep on unrestricted animals in the wild using a recently developed miniaturized EEG recorder and found that brown-throated three-toed sloths (*Bradypus variegatus*) inhabiting the canopy of a tropical rainforest only slept 9.63 hours a day, over 6 hours less than that reported in captivity. Although the influence of factors such as the age of the animals cannot be ruled out, our results suggest that sleep in the wild may be markedly different from that in captivity.

Before this, scientists had generally believed that sloths slept for between eighteen and twenty hours a day. In fact, as Voirin later explained, 'We have shown that in the wild, they sleep around nine hours... This may be due to animals being bored in captivity, increased predation risk in the wild, the need to find forage, or one of many other variables. By doing the research in the wild, we have shown sloths are not actually as sleepy as we previously thought.'

The results were quickly picked up by other outlets in the science media. 'Wild sloths are no sleepyheads after all,' proclaimed the *New Scientist* in the headline of a piece written by Jo Marchant. 'Far from deserving their lazy reputation, wild sloths sleep far less than biologists had thought,' Marchant wrote. 'As well as helping repair the sloth's reputation, the result could have profound implications for research into the function of sleep in both animals and humans.' She also quoted a comment by lead researcher Niels Rattenborg which reinforced the view that the sloth is, if not nearly as sleepy as we have thought, still a splendidly docile creature. Speaking of the easy acceptance of the EEG bowler hats, he said: 'They don't seem to be bothered by us putting things on their heads.'

The science newsletter *LiveScience* was also impressed by the research, declaring in its headline, 'Sloths Are Not Total Sloths', while the esteemed science journal *Nature* chose 'Sloths not so slothful' as its headline, explaining that 'In the wild, sloths are no lazier than the average teenager.'

Above all, however, this research finally showed that Dr James F. Toole had been asking the wrong question in 1971 when his paper 'Why Are Sloths so Slothful?' was published in the *Transactions of the American Clinical and Climatological Association*. His studies of muscle movements in sloths had been designed to discover whether they suffered from myotonia, which he described as 'an abnormality of skeletal muscle in which movement is slow in performance and delayed in cessation'. His conclusion, however, was that 'we cannot verify our initial suspicion that the sloth might be myotonic'. What Rattenborg and Voirin and those who have followed them more than thirty-five years later have shown is that the sloth might not even be slothful.

Even as long ago as 1939, the question of the slothfulness of sloths was raised in the journal *Science* by S.W. Britton and R.F. Kline of the Physiological Laboratory of the University of Virginia. Their brief communication was entitled 'On Deslothing the Sloth' and began with a fascinating thought: 'During several visits to Panama and while making other observations, the possibility of raising the level of activity of the sloth made an interesting appeal.' After explaining what an excellent subject the sloth makes for an academic study into the effects of various substances or conditions, they report several methods that were found to increase the level of a sloth's activity:

> Mere exposure to the tropical sun for an hour or two raised the rectal temperature 4° or 5°, and thereupon the activity of the animal became much greater. This was evidenced by its rate of travel along the under side of a twelve-foot horizontal pole, timed by stop-watch. Again setting up an emotional reaction in the sloth, by simple feints and passes before it, augmented its speed very markedly.

This rather contradicts the findings of other researchers, including William Beebe, who said that even firing a gun next to a sloth seemed to have no effect on the animal – not because of deafness, but because the animal was not in the least interested.

Britton and Kline do, however, say that:

Raising the body temperature appeared to be the best stimulator; on the average the increments in rate of walking on warming approximated 50 per cent, and several cases showed increases of over 100 per cent... It appeared from several hundred tests that the two-toed sloth averaged a little over three hours to the mile and three-toed animals almost four-and-a-half hours. The slowest individual tardigrades, however, took over six hours for the distance. Under excitation such as that noted above [which included speed increases after the sloth had been treated with drugs such as adrenalin], the mile was possible in about two hours, and in a burst of speed by one animal only, a mile an hour was accomplished.

The average figures they quote, however, are about twice as fast as the modern estimates of between two and four metres per minute.

This rather wide range in estimates of sloth speed make it difficult to be sure of the results when trying to compare a sloth with other animals in that respect. Most estimates, however, put a sloth clearly faster than a slug or snail, but whether a sloth would beat a tortoise in a race is not so clear. Noticing that two-toed sloths and tortoises shared the same rainforest enclosure in London Zoo, I asked which was faster, and a keeper told me that he had witnessed what happened when a sloth met a tortoise walking towards each other along the ground. Apparently this results in a rather bad-tempered stand-off, with both animals refusing to give ground. A formal race, however, has yet to be held.

To appreciate the slothful nature of sloths properly, we have to go back a few tens of millions of years to the time when a few sloths, with amazing foresight, or perhaps a great deal of luck, decided to go a different way from the giant ground sloths. That way was upwards. Instead of getting bigger and bigger, and defeating all their enemies by grabbing whatever food they wanted by virtue of their impressive bulk, this small minority of sloths were the pacifists of their age: they decided to live up trees, not eat meat, and stay out of the way of predators.

Curiously, some recent research suggests that giant sloths were vegetarians too. In 2017, a team of scientists from the University of Tübingen in Germany employed a method involving measuring levels of carbon isotopes in fossilized megatherium bones to determine the likely ratio of protein and mineral content in their diets. By comparing the results with similar ratios in the bones of both modern mammals and fossils of animals whose diets are better known, they reached the conclusion that the megatherium was a vegetarian. Its vast bulk was therefore not used to catch and devour prey, which just adds to the mystery of why they, and other giant sloths, died out, while the small ones lived on.

The reasons modern sloths now seem so weird, so vulnerable, and so lazy may be the reasons they have survived. As modern research has shown, everything that has puzzled us about sloths for centuries can be explained by their evolutionary quest for energy efficiency. In short, they have become masters of an alternative lifestyle.

Some time ago, I saw a fridge magnet showing a picture of a sloth captioned with the words 'I'm not lazy, I'm energy-efficient'. I immediately adopted that as a personal slogan, and frequently, when I don't feel like exerting myself, say, 'I'm feeling a bit energy-efficient: I think I'll have a nap.' In the sloth's case, however, it's true, as we shall now see by looking at some aspects of its physiology and behaviour. Sloths are arboreal folivores, which means that they live in trees and eat leaves. So let us see how well designed they are for that purpose. We shall go into each of the following items in greater depth later, but for the time being, this should serve as a summary:

1. They have incredibly slow digestion – which enables them to extract every bit of energy from leaves, which are notoriously hard to digest and have very low energy content.
2. Eating leaves on the trees they inhabit means that they do not need to move much.
3. Staying motionless at the top of trees keeps them away from ground-level predators.
4. They only go to the toilet once a week, which further helps all of the above.
5. Their motionlessness encourages the growth of algae in their fur which acts as camouflage and makes them hard to detect by airborne predators such as eagles and owls.
6. They are contenders for the title of animal with the slowest metabolism, which also helps their energy efficiency.
7. Rather than living on top of branches like most tree-dwellers, they spend a great deal of time hanging from them, which has been shown to require less energy.
8. Their claws are ideally designed to clamp over branches and hang from them, again with minimal energy expenditure.
9. They can go for twenty minutes or more without breathing, then resume without even panting.
10. Unique among warm-blooded creatures, their blood temperature adjusts over a wide range to match that of their

surroundings thus saving the energy that would be needed to maintain a constant temperature.

The last of these is so surprising that the twenty-page opening chapter of M. Goffart's classic treatise *Function and Form in the Sloth*, written in 1971, concentrated entirely on the subject of thermoregulation. To judge from the number of research papers he mentions on this topic, one may see the importance ascribed to it by early sloth researchers. One might also be surprised at how many scientists seem to have spent much of their time inserting thermometers into the bottoms of sloths, but that is another matter. Before leaving it, however, I should mention the opening sentence of a paper entitled 'Poikilothermism in the Sloth' (by S.W. Britton and W.E. Atkinson, *Journal of Mammalogy*, 1938): 'During the course of studies on renal and adrenal function... opportunity was taken to secure readings of rectal temperatures of sloths.' Evidently this is not the sort of opportunity diligent researchers will forgo.

From a scientific viewpoint, however, this was perhaps the starting point for establishing the energy efficiency of sloths. As Becky Cliffe, perhaps the greatest pioneer of the modern sloth revolution, puts it: 'They are energy-saving mammals taking life at a slow pace to avoid the rush and tumble for food, while subscribing the movement patterns that help them avoid being identified as prey.' And she then added: 'There must be a lesson somewhere in that for all of us.'

Artist's impression of a giant sloth – needing a recount of its number of toes.

Chapter 6

ANATOMY

And then came the three-toed sloth. Stupid sloth.
It was a crazy-looking beastie, all arms and
bristling grey fur; its body was a blob, the kind
of shape a six-year-old would draw for a pig,
and its face was flattened like a raccoon that
had run full tilt into a brick wall.

Tony James Slater, *That Bear Ate My Pants!* (2011)

Long arms, short legs, sharp claws, a shaggy coat and an endearing smile is all most of us see when we look at a sloth, but the animal's anatomy possesses far more unusual features which, when taken together, make it unlike any other creature.

Let's start with the teeth. Sloths are edentates, which ought to mean that they have no teeth. The name in their case, however, refers only to their lack of front teeth. Both two-toed and three-toed sloths have nine back teeth on each side of the jaw, five on the upper jaw, four on the lower. While the two most central of the upper teeth can be used for biting, the remainder have clearly evolved for chewing and grinding. This tooth formation has also been seen in the fossils of giant sloths, so must have occurred long ago in their evolution. Unlike most mammals, their teeth are not covered with enamel. They do not have any deciduous teeth (milk teeth in the young that fall out as they reach adulthood) but have a single set of teeth that continue growing and being worn away throughout their life. Having had tens of millions of years to evolve the right balance, the rate of growth matches the rate at which the teeth are worn away.

Another thing that matches their dental formation is the sloth's long arms. With leaves from the upper branches of trees being their favourite food, the most energy-efficient feeding method is not to chew the leaves off the branches, which would involve a good deal of movement, but to rip them from the tree with their claws, move them into the mouth, chew and swallow. This facilitates their ability to pick their favourite tender young leaves from the highest branches which would not support the weight of the sloth. When it does get close enough to the leaves, it can tear them off with its tough lips before chewing. Their favoured position for eating, incidentally, is upside down, with their tongue used to aid swallowing by continuously pushing the food backwards.

After chewing and swallowing comes digestion, which is another unusual aspect of the sloth's design. The trouble is that leaves consist mainly of cellulose and water, neither of which are rich in nutrients. The cellulose has to be broken down by micro-organisms in the sloth's stomach into fatty acids, and this takes time. A long time, in fact, which helps explain why the sloth has a four-chambered stomach rather like that of ruminants such as cows or deer. The food may take over a month to pass through all these stomachs, enabling all the energy to be extracted. As a result, the sloth's stomachs are generally rather full. Compared with other herbivores, the sloth does not eat much, but its stomach contents, plus digestive organs, may account for half of its total weight or more, and when the sloth comes down its tree once a week to poo and pee, it can expel a quarter of its body weight. Having disposed of the sloth's waste products, let us move on to its bones.

As we have mentioned before, the sloth, along with anteaters and armadillos, is a member of the mammalian superorder Xenarthra, meaning 'strange joints', and it well deserves that name, which refers to the extra articulations connecting the bones at the base of the spine. In anteaters and armadillos, this seems to support their digging abilities; in sloths their xenarthran tendencies are what enables them to hold on with their hind legs while apparently dislocating their legs at the hip to incline their bodies at an acrobatic angle of 90 degrees.

As well as the strange leg attachments, however, sloths also have extraordinary necks. Other mammals, from the long-necked giraffe to the mole (which appears to have no neck at all), almost always have seven cervical vertebrae. Remarkably, the three-toed sloth is usually described as having nine, while the two-toed has an average of six. Actually, a two-toed sloth's neck may have between five and seven neck bones, while in the case of the three-toed variety, I say 'usually described' because there is some argument as to the true nature of these bones. In 2010, a team from Cambridge University published a paper in the *Proceedings of the National Academy of Sciences of the United States of America* entitled 'Skeletal Development in Sloths and the Evolution of Mammalian Vertebral Patterning', which showed that the extra lower bones in the neck develop before the neck bones themselves. They concluded, as co-author Dr Robert Asher put it, that 'those extra few are actually rib cage vertebrae masquerading as neck vertebrae'. I think this confirms the conclusion of Buchholtz and Stepien in their 2009 paper 'Anatomical Transformation in Mammals: Developmental Origin of Aberrant Cervical Anatomy in Tree Sloths' (*Evolution & Development*, 11:1, 69–79), where they say: 'We identify global homeotic repatterning of abaxial relative to primaxial mesodermal derivatives as the origin of the anomalous cervical counts of tree sloths',* but it is worth mentioning that it's the extra bones in the neck that give the three-toed sloth the remarkable ability to rotate its head through 270 degrees in either direction. In other words, it can twist its head from forward-looking to looking behind itself, then twist through an extra 90 degrees – the sort of rotation any owl would be proud of.

Interestingly, the number of neck bones is just about the major thing on which two- and three-toed sloths diverge (apart from toe numbers, of course). If we go back 35 million years or so, we find ourselves in a prehistoric world inhabited by giant ground sloths and small tree sloths in which the two- and three-toed varieties are about to go their separate evolutionary ways. This is almost twice as long a period of time as humans and orang-utans have had to become different creatures (and *Homo sapiens* has only been around for the last 300,000 years or so). Humans and apes and monkeys, however, have evolved along very different paths. By contrast, however, in general appearance and habits, the two types of sloth have made much the same choices: their current appearance is similar and they both hang upside down in trees a lot. But while a couple of pairs of

---

* This sounds so much more scientific than saying 'the extra neck bones are ribs really', doesn't it?

ribs in the three-toed have become extra neck bones, the two-toed have gone in the other direction, reducing their neck bones by one or two. Quite why this should have happened is as great a mystery as the difference in toe numbers. Three toes good, two toes also good, as George Orwell might have put it if there had been any sloths in his *Animal Farm.*

S.W. Britton, in his classic 1941 paper 'Form and Function in the Sloth' (not to be confused, of course, with M. Goffart's equally classic 1971 book *Function and Form in the Sloth*), suggests that both two- and three-toed sloths developed from a five-toed creature. '*Bradypus* is said to have lost the first and fifth toes and *Choloepus* the fourth also in the forelimbs.' However, he goes on to mention the case of a polydactylic sloth (i.e. one that has more digits than usual) 'in which all five claws were present in the right hind-limb of *Bradypus*.'

In 2015, Becky Cliffe, in her excellent blog from the Sloth Sanctuary in Costa Rica, also mentions finding a polydactylic sloth:

> He was in perfect health, but there was something strange going on. This three-fingered sloth had four toes. Four perfectly formed toes on his left foot. We often see sloths with missing digits due to genetic deformities, but this is the first time we have ever seen a sloth with extra toes!

She says they named him Quatro. He was a master of hide-and-seek and was the most difficult sloth she worked with.

Nor is sloth fur to be left out of the animal's general claims to weirdness, for it has two features that are unusual and particularly well designed for its lifestyle, especially when it is raining. Unlike most quadrupeds, whose hair parts in the middle of their back and grows towards the stomach, the sloth's hair parts down the length of its stomach and grows towards the back. When you think about it, this makes great sense when hanging upside down in the rain: the water just flows down in the direction of the fur and drips off its back. The second unusual feature of sloth fur may also aid drainage: the fur consists of two layers of hair, a soft undercoat and a thicker, coarser coat above it, but the hairs of the upper coat have cracks and grooves that soak up water like a sponge and keep the lower coat dry. This feature also makes the upper layer of the coat ideal for the algae that grow in the sloth's fur which is responsible for the greenish look of sloths in the wild that helps to camouflage them against the attention of eagles.

The sponge-like tendency provided by these cracks and channels in its outer fur may also explain what led to one of the most amusing and delightful photographs in Hermann Tirler's 1966 book *A Sloth in the Family*. The picture illustrates a passage in which Tirler explains his wife's reluctance to have a wet sloth in the house despite her otherwise very positive attitude to their unusual pet, and shows her solution to the soggy-sloth dilemma: before being allowed back into the house after a storm, it is shown hanging on a laundry line to dry. A sloth may be the most drip-dry of all pets.

Finally, before leaving the sloth's anatomy, we should say something about their limbs and claws. Most quadrupeds have four legs that support their bodies as they walk around. Humans walk around on two legs, using the other two limbs as arms. Sloths, however, come the nearest to having four arms, all of which serve the function of grasping. Instead of supporting their bodies, they use their limbs to secure themselves on branches as their bodies hang below.

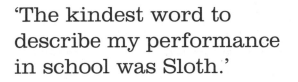

'The kindest word to
describe my performance
in school was Sloth.'

Harrison Ford

The muscles in any creature's body may be divided into 'retractors', which hold and cling to things and pull things towards the animal, and 'extensors', which give pushing power. For legs to support the body and enable walking and running, they need extensor muscles. For most functions, however, the strength in our arms comes from retractors. While sloths have only half the overall muscle weight of animals of comparable size, their muscles are primarily retractors, which gives them tremendous grip. Even a baby sloth gripping its mother's fur can hardly be prised off without ripping the fur itself, while native South Americans have known for generations that the best way to capture a sloth is not to try to pull it from a tree but to cut the branch it is resting under and carry off the branch itself. Despite the strength of its grip, the sloth expends very little energy in supporting its own weight in that manner, or moving along, underneath a branch, by changing its grip by moving the arms.

On the ground, however, it is not equipped for movement at all. Its legs cannot support its weight and it is reduced to dragging itself along by finding something to wrap its claws around or dig them into. On a perfectly smooth surface, they are completely helpless.

The surgeon, anatomist and theological philosopher Sir Charles Bell (after whom Bell's palsy was named) wrote in 1833 in his treatise *The Hand: Its Mechanism and Vital Endowments as Evincing Design*:

> Modern travellers express their pity for these animals: whilst other quadrupeds, they say, range in boundless wilds, the sloth hangs suspended by his strong arms – a poor, ill-formed creature, deficient as well as deformed, his hind legs too short, and his hair like withered grass; his looks, motions, and cries conspire to excite pity; and, as if this were not enough, they say that his moaning makes the tiger relent and turn away. This is not a true picture: the sloth cannot walk like quadrupeds, but he stretches out his strong arms, – and if he can hook on his claws to the inequalities of the ground, he drags himself along. This is the condition which authorizes such an expression as 'the bungles and faulty composition of the sloth'. But when he reaches the branch or the rough bark of a tree, his progress is rapid; he climbs hand over head, along the branches till they touch, and thus from bough to bough, and from tree to tree; he is most alive in the storm, and when the wind blows, and the trees stoop, and the branches wave and meet, he is then upon the march.

We move on to a recent highly significant discovery about sloth anatomy, which provided the answer to a question of great importance that few had even thought to ask: namely, why doesn't the huge weight of material being digested in a sloth's stomach weigh down on its vital organs, particularly its lungs, and crush the life out of it or at least make breathing more difficult and energy-consuming?

In a paper entitled 'Mitigating the Squash Effect: Sloths Breathe

Busy doing nothing

How could you call me a deadly sin?

The joy of hanging upside down in the wind

In the wild in Costa Rica

A male three-toed sloth in Costa Rica proudly displays the bright orange patch on its back

On its way down for the weekly toilet appointment

Three legs hanging, one arm foraging

Apart from hiding, a sloth's claws are its only means of defence

Three legs hanging, one arm eating

Getting to know the ropes

The wrong way to take a selfie

A characteristic welcoming smile from a three-toed sloth

Easily Upside Down' in the journal *Biology Letters* in 2014, a team led by Becky Cliffe reported that three-toed sloths 'possess unique fibrinous adhesions that anchor the abdominal organs, particularly the liver and glandular stomach, to the lower ribs'. These adhesions, which Cliffe in one report likened to coat hangers, are ideally placed to support the weight of the stomach and bowels while the sloth is upside down and prevent the lungs from being squashed. They also contribute to the sloth's remarkable performance in energy-saving. 'We estimate that these adhesions could reduce a sloth's energy expenditure by 7 to 13 per cent when hanging upside down,' she commented, adding that work is still being done on a 'whole array of fascinating anatomical features', such as those that help the sloth eat upside down and stop the blood rushing to its head.

We must finally mention an anatomy-related matter that has been the subject of two recent academic studies: the question of whether sloths are predominantly left- or right-handed. Most recently, a paper by Shery Duque Pinheiro and Carlos Esbérard, entitled 'Laterality Evaluation as an Evidence of Brain Asymmetry in Sloths *Bradypus variegatus*', was delivered at a Neuroscience Symposium in Brazil in 2012. It was based on experiments involving twenty-three brown-throated sloths taken from two populations living in different areas. The sloths were placed in an area with two timber supports allowing them to move horizontally or vertically. When they moved, note was taken of whether they initiated the movement with a right or left limb. They report:

> Comparison between the two areas' populations showed different patterns. The population of area 1 showed a significant difference in the use of limbs in both the vertical task

and the horizontal, with a preference of the right hand/limb
to perform the movements. The population of area 2 showed
no significant preference for either limb, both in horizontal
or vertical displacement.

In other words, the sloths in one group were predominantly right-
handed, but those in the other group showed no signs of either
right- or left-handedness. They very reasonably concluded that
'More research is needed.'

Earlier research on handedness in sloths, also by Pinheiro and
likewise conducted in Brazil, had appeared in 2009 under the title
'Study of Laterality in *Bradypus variegatus* on Feeding Behaviour'.
Both papers were based on research conducted as part of the Sloth
Project at the Centenary Park in Barra Mansa, and this one had
been concerned in particular with which of their hands sloths use
for eating and other activities. Taken altogether, she reported signifi-
cantly more use of the right forelimb, but when it came to studying
the differences between individuals, 'six sloths were considered
right-handed and six left-handed', further reporting that only one
individual showed consistent preference for the left forelimb, which
suggests that left-handed sloths generally use their right hands more
than right-handed sloths use their left. 'The sloth population did not
present hand preference in feeding behaviour,' we read.

This is all very confusing. Some are right-handed, some are
left-handed, but they are mostly ambidextrous when it comes to
eating, and left-handed sloths use their right hands quite a lot
anyway. Further research is clearly still needed.

Chapter 7

SEX AND THE SOLITARY SLOTH

**'You know, sloth is a sin,' he says softly.
'I prefer to think of it as an adorable animal.'**

Ella James, *Sloth* (2015)

In 2013, London Zoo decided it was time for their two-toed sloth, Marilyn, to get a boyfriend. The zoo had never successfully bred a baby sloth before, but they sensed that Marilyn might be up for it. So they flew over from Germany a boy sloth named Leander and, more in hope than expectation, waited to see what would happen.

In their natural habitat, sloths are highly solitary creatures. Three-toed sloths keep themselves to themselves, hanging around in their own tree with no apparent need or desire to communicate with other sloths. Their two-toed cousins sometimes live two to a tree, but that is still the exception, with single-sloth occupancy generally the preferred tree-dwelling manner. In either case, when a boy sloth meets a girl sloth, it can take several months before he invites her back to his place.

The same slow progress has been seen in sloth romances in other zoos, with boy–girl pairs studiously ignoring each other for months. This was certainly the pattern followed by Marilyn and Leander. After the sloths had been together for over six months, the keepers were astonished to discover that Marilyn was pregnant. As one was quoted as saying, 'We did not even know they had acknowledged each other's presence.'

Marilyn gave birth in May 2014 and, factoring in the expected eleven-month gestation period of the two-toed sloth, it was calculated that conception must have happened six months after Leander's arrival, which is thought to be pretty quick going for a pair of sloths.

In the wild, accounts of sloth lovemaking are very rare. In 1926, the great naturalist William Beebe, who was one of the first to spend a good deal of time observing sloths in their natural habitat, wrote: 'I have watched two courtships, one of an immature male, and the other of an animal of full size and colour. Both were alike in their absolute directness and simplicity.' He went on to describe how the males climbed up to the female and tried to grab her: 'In the first instance, where the female had a month-old baby clinging to her fur, she lunged leisurely with full force at the disturber of her peace. The other female simply mounted higher, and when she could ascend no more, she climbed down and across her suitor, leaving him stranded on the lofty branch looking vaguely about.' The male sloth then tried to reach out for an iguana lizard, which he thought might be his beloved, before getting back to his pursuit of the female sloth. 'This unemotional pursuit continued for an hour, when he gave up for good and went to sleep.'

Accounts of successful sloth mating in the wild are also very sporadic and provide insufficient evidence to build a reliable picture. Brazilian researchers have reported seeing sloths mating in two positions, with the male mounting the female from behind, or face-to-face. Unlike most of their other activities, lovemaking seems generally to be brief, lasting only a few minutes, but the male may then take a short rest and try again. On the rare occasions that researchers have observed sloths mating, they have tended to sense the unusualness of the occasion and written it up in great detail for the benefit of others. What follows should therefore perhaps carry a parental advisory warning.

In 2008, a team of Brazilian researchers published a paper entitled 'First Observation on Mating and Reproductive Seasonality in Maned Sloths *Bradypus torquatus* (Pilosa: Bradypodidae)'. Coming straight to the point, they spare no blushes by saying: 'A pair of maned sloths was observed copulating in September 2005 in the Atlantic Forest region of south-eastern Brazil.' After a good deal of background information on gestation periods and possible seasonal mating behaviour of sloths, they return to the X-rated business:

> The two were high in the canopy and so tightly embraced that initially we thought there was only a single large sloth in the tree. For this reason we could neither discern copulation movements nor see if their relative position was ventral-ventral or ventral-dorsal... The two adults stayed embraced for approximately 7 min and then separated from each other, remaining side by side on the same tree trunk.

The female sloth, incidentally, was nursing a baby while all this was going on. It was also pouring with rain, which could be seen as adding drama to the intense cuddling. The researchers report that when it stopped raining several hours later, they were able to climb the tree and capture the male to confirm that it was indeed sexually active and 'had a prominent penis'.

Tabulating the known dates of sloth copulations and births, the authors say that 'the little information that does exist indicates some reproductive seasonality', but suggest that not all sloths, even those from the same area, fall into the same pattern. To support this, they mention earlier research involving a female kept together with a male in semi-captivity who was observed copulating with another male that entered her enclosure, 'while the male who lived with her did not attempt to copulate nor did he show any sort of reaction'.

As far as I have been able to discover, the academic literature contains no reference to homosexuality in sloths, though my searches have revealed one appalling riddle:

Q: What do you call a gay sloth?
A: A slo-mo-sexual.

Meanwhile, back with the heterosexual sloths, two recent research papers bring some surprising revelations.

In December 2012, the online science journal *PLOS One* published a paper, 'Unexpected Strong Polygyny in the Brown-Throated Three-Toed Sloth', which revealed that male sloths are far more promiscuous than one would expect from such a sedentary creature. The research involved taking DNA samples from all the sloths in a given area over the course of two years in order to determine the parentage of the babies. The surprising result was that 'only 25% of all resident adult males sired offspring and one

'Jack, you have debauched
my sloth.'

Patrick O'Brian, *HMS Surprise* (1973)

individual sired half of all sampled juveniles'. In fact, of the nineteen
males and twenty juveniles whose DNA was successfully sampled,
one male was the father of ten children, and another four males
accounted for all the others, being the fathers of four, three, two and
one offspring respectively. Fourteen of the males were found not to
be responsible for any of the children.

Nearly five years later, however, female sloths hit back against
this picture of sexually rampant males. In 2017, the *Journal of
Mammalogy* published a paper, 'Individual Reproductive Strategies
Shape the Mating System of Tree Sloths' (by Mario F. Garcés-
Restrepo and others), showing that sexual promiscuity among
sloths is not the sole preserve of the menfolk. The research took
place on a cacao plantation (sloths love chocolate, apparently),
a tropical forest and a cattle pasture in Costa Rica, and involved
capturing almost 400 sloths and fitting them with radio-collars that
allowed very accurate tracking of their locations. In particular, they
could determine when two sloths were spending time together, and
the results suggested that the females were no sexual slouches. Just
as some males had displayed a high degree of polygyny (multiple
female mating partners), the women displayed polyandry (many

males). 'Female sloths often mated with multiple males over their lifetimes,' they report; 'indeed 70% of *Bradypus variegatus* and 50% of *Choloepus hoffmanni* switched mates at least once during the study... Our study, which spanned 4 breeding cycles in both species, suggests that mating with more than 1 male over time is common in sloths, and greater than previously believed.' (To be fair to the female sloths, we should point out that after mating, the males usually wander off, and after childbirth will play no part in looking after the offspring.)

Courtship is another area which is only just beginning to be understood. Some say that females climb to the tops of trees, then emit a whistling sound to attract males. The sound only lasts for a second but may be repeated several times. This whistling cry is said by many to account for the onomatopoeic name of '*ai*' being given to the three-toed sloth, but it is unclear whether it is specifically a mating call.

William Beebe reported that the pitch of the whistling note was 'not far from the upper limit of human whistling' and was always precisely the D sharp above middle C. 'A most interesting thing is the way their hearing is exactly attuned to this note,' he wrote, giving details of an experiment in which he placed two mother sloths in a cage, took away their offspring, then went sixty feet away and whistled at them. 'Slowly but surely both heads turned in my direction and a male, high up on his tree, also turned at the same instant,' but only when he whistled a D sharp. Other notes produced no reaction whatsoever. An anxious mother might show interest in a note a semitone away, but even a D natural had no effect on most sloths, 'while D sharp aroused all the interest which their poor, dull minds could bring to bear'. Beebe does not, however, mention this

D-sharp whistle specifically as part of courtship, so whether girl sloths whistle at boy sloths remains a matter for speculation.

Another, much more recent, theory of sloth courtship involves mutual poo-sniffing and has been advocated as a good reason for three-toed sloths climbing down their trees once a week always to poo in the same place. Quite how sniffing poo at the foot of a tree fits in with whistling from its top is another question that needs answering. We shall leave the sloth's curious toilet habits and its possible link to mating behaviour until Chapter 9.

After the mysteries of courtship and mating come gestation and childbirth, and the first of these is a bit of a mystery too. The trouble is that sloths, even in captivity, have never been sufficiently closely monitored to know precisely when pregnancy has occurred and, when it does occur, detectable signs may not be seen for some time. Estimates of the length of a sloth's gestation period therefore vary quite widely. One authority asserts that all sloths have a gestation period of 'about six months'; another estimates ten. Four to six months for species of three-toed sloth, but ten to twelve months for two-toed, say some, while yet others insist it is five to six months for Linnaeus's two-toed sloth but eleven to twelve for Hoffmann's two-toed variety. All they seem agreed on is that female sloths give birth to one baby at a time. The most convincing recent research into sloth gestation periods came in a study in Germany in 2016 which used ultrasound to detect early signs of pregnancy with greater reliability than earlier attempts to do so. Their conclusion was that the gestation period of both species of two-toed sloths is between 330 and 350 days.

Finally, some news to report that is both good and bad: the first ever baby sloth delivered by Caesarean section was born at the

Sloth Institute in Costa Rica in October 2014. The baby's heavily pregnant mother had been brought there after falling from a tree and suffering head injuries. An emergency C-section was performed after the mother failed to respond to treatment and a baby boy sloth was delivered successfully. Sadly, the baby was found to have a heart murmur and died a week later. The mother died the following day from brain injuries suffered in her fall.

Chapter 8

THE DEADLY SIN

You must avoid sloth, that wicked siren.

Horace, *Satires* (c. 30 BC)

Having established that sloths are not slothful but merely supreme examples of energy efficiency, it is time to deal with the so-called Deadly Sin of Sloth.

The word 'sloth' itself does not occur in the King James Bible, but there are seventeen instances of 'slothful' or 'slothfulness' (of which twelve feature in the Book of Proverbs), and none of those is complimentary. When we get to the New Testament, admonition towards slothfulness becomes more severe. Matthew 25:26 refers to a 'wicked and slothful servant', while St Paul, in his epistle to the Romans (12:11) advises them to be 'Not slothful in business' but 'fervent in spirit; serving the Lord', while he also comes straight out to the Hebrews (6:12), advising them 'That ye be not slothful'. The Book of Ecclesiasticus in the Apocrypha (22:1–2) comes down most heavily of all on sloth, telling us that 'A slothful man is compared to a filthy stone, and every one will hiss him out to his disgrace. A slothful man is compared to the filth of a dunghill; every man that takes it up will shake his hand.' (Presumably this handshake is to shake off the dung rather than by way of introduction.) We need to delve rather deeper into the history of the relationship of the Christian Church to sloth, however, to understand how it became a Deadly Sin.

The Seven Deadly Sins owe their origin to the writings of the fourth-century monk Evagrius of Pontius (345–99), also known as Evagrius the Solitary. While pursuing an ecclesiastical career at Constantinople, where he became a deacon, then archdeacon, Evagrius fell in love with a married woman but is said to have been saved by having a vision in which he was imprisoned by soldiers at the request of the woman's husband. Saved from such a fate, he put his near-disaster down to a combination of vanity and lust and set about drawing up a list of evil thoughts or temptations. He ended up with a list of eight such sins which, in order of seriousness, were: Gluttony, Lust, Avarice, Sadness, Anger, Acedia, Vainglory and Pride. These were the temptations that Evagrius saw monks facing that might lure them away from their religious duties and studies.

The nearest to Sloth in this list is Acedia (also known as Accidie), which was the term used for a sort of spiritual listlessness, a mixture of apathy, boredom, laziness and a rejection of God's divine message. Evagrius's list underwent various modifications under several hands over the next couple of centuries, but Acedia remained one of the 'Evil Temptations', or 'Capital Vices', as they were generally called until 590 when Pope Gregory (St Gregory, or Gregory the Great) revised the list by combining Vainglory (a sort of unjustified vanity) with Pride, Acedia with Sadness, and adding Envy to bring the list back up to seven, which are more or less the Deadly Sins we know today.

The precise nature of the sin of Sloth remained a matter of theological debate for centuries until it was considerably clarified by Thomas Aquinas in the thirteenth century in his *Summa Theologica*. 'Sloth,' he wrote, 'is an oppressive sorrow which so weighs upon man's mind that he wants to do nothing.' It is, he goes on, 'a sluggishness of the mind which neglects to begin good.' The problem with Sloth, as he had already made clear, is that, unlike the other sins, it may be seen as a sin of omission rather than commission, but Gregory was in no doubt that Sloth was just as evil and sinful as the others, for 'sloth seeks undue rest in so far as it spurns the divine good.' Evagrius, too, had put the slothful sin of Acedia in a class of its own. It was a monk's duty to overcome the other sins, he said, but overcoming Acedia was the only way to get closer to God.

Aquinas, of course, wrote his *Summa Theologica* in Latin, and the word he used for Sloth was *acedia*, which, as we have seen, meant more than just laziness. The word 'sloth' (or slouth, slouthe, slauthe), meaning slowness or laziness, was already known in Middle English and the earliest citation given in the *Oxford English Dictionary* dates back to the Lambeth Homilies, written around 1175, but it took a long time for Sloth to become a Deadly Sin in English by that name alone.

At first, Sloth was listed by several authors as one of the components of Accidie, but by the early seventeenth century it was being used as the name of the Deadly Sin itself. One of the earliest such references came in a book entitled *The Seven Deadlie Sinns of London: Drawne in seven several coaches, Through the seven severall Gates of the Citie* (1606) by the English dramatist and pamphleteer Thomas Dekker. The book lists how each of the seven sins was

responsible for bringing plague to London, and when he reaches
the sin of Sloth, he says this:

> How then dares this nastie, and loathsome sin of Sloth
> venture into a Citie amongst so many people? who doth he
> hope wil giue him entertainment ? what lodging (thinks he)
> can be taine vp, where he and his heauy-headed company
> may take their afternoones nap soundly?

The sin was therefore already being called 'Sloth' in English before
the animal acquired that name, and the early writings about the
animal can hardly have failed to be influenced by the 'nastie, and
loathsome' reputation of the Deadly Sin.

Curiously, however, just as the English-speaking world was
beginning to incline towards calling the animal a 'sloth', the Church
was tending to refer to the sin as *pigritia* or *ignavia* (both taken
from Latin words for 'indolence'). To one writer at least, the rather
pleasant human weakness of sloth was far too nice a word to give
to a Deadly Sin, let alone a much-loved animal. Evelyn Waugh
summed this up in a 1962 essay on Sloth:

> The word 'Sloth' [Latin *acedia* or *accidia*] is seldom on
> modern lips. When used, it is a mildly facetious variant
> of 'indolence' [Latin *pigritia* or *ignavia*], and, surely, so far
> from being a deadly sin, [such indolence] is one of the most
> amiable of weaknesses... How then has Sloth found a place
> with its six odious companions as one of the Mortal Sins
> [Latin *Peccata Mortalia* or *Vitia Capitalia*]?

Waugh goes on to answer his own question:

> We only know that Hell is for those who deliberately choose it...What then is this Sloth which can merit the extremity of divine punishment?... Sloth is the condition in which a man is fully aware of the proper means of his salvation and refuses to take them because the whole apparatus of salvation fills him with disgust.

Finally, however, and perhaps to redress the anti-sloth bias of all this sinfulness, we should note that the sin of Sloth has in several cultures been seen as worthy of special respect, even to the extent of having its own divine entities.

In Buddhism, for example, sloth-torpor is seen as one of the five 'hindrances', hostile to meditation. It is characterized by lack of energy and opposition to wholesome activity.

False of heart, light of ear, bloody of hand; hog in sloth, fox in stealth, wolf in greediness, dog in madness, lion in prey.

William Shakespeare, *King Lear* (1605)

In ancient demonology Belphegor was the Prince of Hell who tempted mortals through the sin of Sloth, offering them inventions that would make them rich without effort. According to some sixteenth-century demonologists, his power was strongest in April. He would often appear as a naked woman, but also sometimes as a gruesome demon. Belphegor has been associated with the Moabite god Baal-Peor of Old Testament times and, more recently, has become the only ancient demon with his own prime number. 'Belphegor's Prime' is 1000000000000066600000000000001, which was given its name by the American mathematician Clifford Pickover, in view of its remarkable superstitious connotations: the Number of the Beast, 666, from the Book of Revelation right in the middle, surrounded on each side by (unlucky) thirteen zeroes and a 1 at each end. Belphegor himself also appears in Milton's *Paradise Lost*.

For a more genuinely womanly apparition, I can offer Aergia, the Greek goddess of sloth, also known as Ignavia or Pigritia. The daughter of Aether and Gaia, she was said to guard the court of Hypnos in the Underworld.

Meanwhile the Hindus had Jyeshtha, goddess of sloth, who was also goddess of inauspicious things and misfortune, poverty, sorrow, ugliness and crows, so with such a busy and diverse portfolio could not have been all that slothful herself.

Last of all, we should not forget the Bushyansta of Persian Zoroastrianism. They were the yellow demons of lethargy and sloth who caused men to oversleep and miss their spiritual obligations. They are usually depicted with unusually long hands.

Chapter 9

POOPING

There were other signs of removal about,
with queer narrow footprints like those I could
imagine made by a sloth.

H.G. Wells, *The Time Machine* (1895)

Authentic sloth poo at the foot of a tree in Costa Rica.

The toilet habits of the sloth, perhaps more than any other creature, have aroused great fascination. Around once a week (about every seven days for Bradypus, and four to five days for Choloepus) they come down the tree to poo and pee on the ground, after which they climb back up the tree. This, for a sloth, is an astonishingly laborious procedure and can take hours to complete. Compared with the rest of the sloth's lifestyle, it is also highly energy-consuming. Furthermore, it is a very risky endeavour. At ground level, a sloth is very vulnerable to predators. It is not built to run away from any animal sneaking up and surprising it during a toilet break. So why does it adopt this extraordinary ritual?

Part, but not all, of the explanation must lie in the sloth's curious diet and digestive system. The three-toed sloth is very fussy about its food, and generally prefers to eat leaves from the Cecropia tree. Well, they say there's no accounting for taste, but Cecropia leaves have very little going for them in terms of their nutritional value. To make up for this, as we have seen, sloths have four stomachs, which the food must pass through before being excreted. Even the cellulose in the leaves, which for most animals is highly resistant to the digestive process, is broken down, first into sugars, then fatty acids. Sheep, cows and other ruminants also have multiple stomachs, but they at least chew the food before swallowing it. Sloths don't have many teeth, and after a few cursory chews to break up the leaves, they just swallow them and leave the gastric juices to do the work.

The result of this slow process is that an item of food may remain in the sloth's stomach for a month or more, in the course of which the sloth, as might be expected, puts on weight. By the time its next toilet trip comes along, the sloth will be ready to lose over a third of its body weight when it urinates and empties its bowels.

There is some dispute among sloth researchers over whether two-toed sloths are as fussy in these matters as their three-toed cousins. It is clear from many reports that members of the two-toed variety do not always bother to come down to poo and pee at ground level at all, instead performing these functions while still hanging from the tree's uppermost branches, letting their waste matter fall under gravity. It has even been stated that this is the predominant form of behaviour among two-toed sloths. One might have expected it to be the other way round: as two-toed sloths spend more of their lives upside down than three-toed sloths, it would seem to make less sense for them to rely on gravity to take their poo to the ground. Excreting while inverted could be very messy. The once-a-week toilet breaks of all sloths, however, have clearly not yet allowed sufficient cases to be observed to permit a definitive answer to this question. There seems to be general agreement, however, that while three-toed sloths dig a shallow hole with their rudimentary tails, poo in the hole, then cover it with leaves, the two-toed varieties, having only tiny tails, do not bother with digging a hole, and when they have finished just leave the poo uncovered.

None of this, however, is to throw any real light on why the sloth bothers to expend so much energy in making the laborious trip down and up the tree to use their equivalent of an outside lavatory when they could perform the same function en suite, just letting it drop to the ground-level sewage facilities. The best theory anyone could suggest was that the sloth poo deposited at the base of a tree acted as manure, providing nutriment to the tree roots, thus helping preserve the sloth's home. That idea was hard to believe for two reasons: firstly, the sloth's digestive system had already extracted almost all the available nutrition from its food, leaving hard, dry clumps of poo in which even insects showed little interest; and secondly, there was no clear reason why depositing the poo at the base of a tree should have better effect than dropping it from on high.

Nobody came up with any convincing theory to account for this until 2013, when a team led by the great sloth explorer Bryson Voirin co-authored a chapter called 'Why Do Sloths Poop on the Ground?', as part of *Treetops at Risk: Challenges of Global Canopy Ecology and Conservation*, a collection of essays edited by Margaret Lowman, Soubadra Devy and T. Ganesh. Their idea was based on an older suggestion that there was some connection between the sloth's toilet behaviour and the numerous sloth moths[*] that are found living in its fur.

It had been known for some time that the moths laid their eggs in sloth poo which provided the necessary nutriment for the eggs to develop and hatch. The newborn moths would then fly up the tree and take residence in the family home, the fur of the sloth. This was clearly good for the moths, but the new idea was that the sloth's diet of leaves missed certain trace elements, such as iron and sodium, which could be picked up from the earth when they came down to poo. The hypothesis was that the sloth's toilet trips gave it the chance to lick the soil, or scrape some up with its claws to be licked off later, thus making up for the otherwise strictly leaf diet.

Voirin's team made a further curious observation about moth behaviour when sloth pooing was due:

> In the days leading up to a defecation event, increasing numbers of sloth moths congregate around the rear of the sloth. Just prior to [the] actual event, as the sloth climbs down toward the ground, the moths become extremely

---

[*] This, incidentally, is a good reason to pronounce 'sloth' with a short vowel: 'slŏth mŏth' rolls off the tongue much more easily than the inelegant 'slōth mŏth'.

active and swarm around the sloth. Once the sloth dung has been deposited onto the forest floor, female moths fly from the sloth and deposit the eggs in the sloth dung.

They end the paper, however, by pointing out that no rigorous evidence exists to test their hypothesis, 'leaving the canopy sloth's terrestrial toilet behaviours an unsolved mystery of the rainforest'. Within months, however, another team of experienced sloth researchers, led by Jonathan Pauli of the Department of Forest and Wildlife Ecology at the University of Wisconsin, published a paper in the *Proceedings of the Royal Society* entitled 'A Syndrome of Mutualism Reinforces the Lifestyle of a Sloth', which also concentrated on the link between poo and sloth moths, but this time they

> SEBASTIAN: Do so: to ebb
> Hereditary sloth instructs me.
> ANTONIO: O, If you but knew how you
>   the purpose cherish
> Whiles thus you mock it! how, in
>   stripping it,
> You more invest it! Ebbing men, indeed,
> Most often do so near the bottom run
> By their own fear or sloth.
>
> William Shakespeare, *The Tempest* (1611)

brought algae into it as well. The lifestyle of an arboreal herbivore (tree-leaf eater), they point out, is rare among mammals, and sloths combine it with the weekly descents to the toilet, 'which is risky, energetically costly and, until now, inexplicable'. The explanation they offered was a bizarre linkage between sloths, moths and algae.

They quite literally drew up a picture of the sloth-moth-poo cycle, as the paper includes an elegant picture of moths descending from a sloth in a tree and laying eggs in its poo, with more moths emerging from the poo flying back up the tree to infest another sloth, and they go on to explain their theory that the moths living in the sloth fur provide the animal with a source of nitrogen which is necessary for the growth of algae, and the algae are an important nutritional source for the sloth.

This picture, they say, is confirmed by an analysis of the stomach contents of sloths which were found to contain algae; furthermore, the greater the density of moths found in the sloth fur, the greater the amount of algae found on the sloth. Previously, the main benefit to the sloth of having algae all over it was thought to be providing camouflage. In its normal resting position hanging from a branch, the sloth's most dangerous enemy is the harpy eagle, which is known to swoop down on sloths and drag them from their trees with its powerful talons. Owls have also been known to pose a similar threat to sloths. But when you are hanging motionless, covered with algae, even an eagle's excellent sight will probably lead to its ignoring you. The same applies to the exaggeratedly slow movement of sloths. Their languorous progress must help evade the motion detectors of predators.

Now, however, another benefit was hypothesized: not only did the algae provide invisibility cloaks, but the sloth would occasionally lick them off for nutritional reasons. But was this neat idea supported by observational evidence? The moths may indeed have been responsible for the algae, and algae had been found in sloths' stomachs – but was this necessarily the same algae? Sloths, after all, were not known to indulge in a great deal of self-preening, and the idea of them licking the algae off their fur did not convince everyone. There was, after all, a good deal of algae growing on trees. Wouldn't it be easier for the sloth to lick that if it fancied an algae treat, rather than licking off its camouflage?

Doubts about the algae-licking theory were expressed by Becky Cliffe early in 2014 in a post on her beckycliffe.com web site. 'It is certainly an interesting idea,' she wrote. 'Unfortunately it doesn't add up.' While agreeing that the sloths with the most algae tend to have the most moths, she pointed out that sloths do not need to eat algae to survive. In fact, sloths kept in captivity on algae-free diets are just as healthy as algae-eating sloths. Furthermore, she points out that despite having spent a great deal of time watching and following sloths in the wild, 'I have never seen anything that looks remotely like licking.'

Her belief is that the sloth–moth arrangement is one that really benefits only the moths: 'The moths have simply taken advantage of the sloth's bizarre bathroom habits and found themselves a nice niche.' She also draws attention to another aspect of this business that the Wisconsin team fails to explain: the curious fact that wild sloths always poo in the same places at the base of certain selected trees.

'We believe that it is all about communication and reproduction,' Cliffe writes, in explanation of the fact that the pheromones in sloth urine and faeces may be what attracts a mate. This has been described by others as the sloths' equivalent of speed dating. They sniff around at piles of poo on the ground, and if they like the smell, they shin up the tree (at a sloth's pace, of course) and meet the sloth responsible.

Quite why the sniffing sloth bothers to come down the tree just to smell some piles of poo is another matter that needs explaining. But before taking our leave of lavatorial matters, we should perhaps mention another crucial question that has led to some dispute: Do sloths fart?

In their 2017 book *Does It Fart? The Definitive Field Guide to Animal Flatulence*, Nick Caruso and Dani Rabaiotti designate sloths among the animals which do not. And yet scientists had only recently been invited to contribute to an open-access 'Does It Fart?' spreadsheet that, at the latest count, answers that same question for eighty-six species – and the answer given for sloths (though no link was offered for verification) was an unambiguous 'Yes'.

The reason behind this confusion probably lies in the sloth's digestive process. The mass of vegetable matter in its stomachs certainly produces a great deal of methane as it ferments – and this is one of the reasons sloths can swim so well, since their stomach gases serve as a huge flotation device, keeping them buoyant in the water. Such flatulence, however, demands release; but the sloth does so through its mouth, not its anus. So, technically speaking, this is an 'eructation': a methane burp, not a true fart.

Caruso and Rabaiotti suggest that sloths may indeed be the only mammals that do not fart, though they admit that nobody is yet sure about bats.

Chapter 10

THE ABOMINABLE SLOTHMAN, AND OTHER MYTHS

The creature had exactly the mild but repulsive features of a sloth, the same low forehead and slow gestures.

H.G. Wells, *The Island of Dr Moreau* (1896)

Neither the post-Columbian North Americans, whose culture is too recent, nor the Europeans, who got completely the wrong idea about sloths when they first learned of them in the seventeenth century, had much of a chance to build up folk tales and legends about sloths, but the indigenous people of Central and South America made up for it in style. We have already met various goddesses and demons of sloth, such as Aergia, Pigritia, Ignavia and Belphegor (see above, p. 102), but those were Sloth the sin, not sloth the animal. For a more down-to-earth mythical sloth, we have to visit South America in search of the Mapinguari.

The legend of the Mapinguari goes back many generations, and features sightings of a creature up to nine feet high resembling a giant sloth or huge ape. Brazil's answer to the North American Bigfoot or Nepalese Yeti, the Mapinguari (the name means 'roaring animal', or possibly 'fetid beast') has always been described as large and hairy with huge claws, but is sometimes depicted with variations such as having only one eye in the centre of its face, or an extra mouth in the middle of its stomach. It also is said to emit a particularly foul odour. According to one Brazilian legend, Mapinguari was originally a mystic who discovered the secret of immortality and was punished by the gods by being turned into a wandering beast.

Alleged sightings deep in the Amazon forest, however, have continued throughout the years, and generally suggest a creature reminiscent of the megatherium ground sloth. Since humans and ground sloths are known to have co-existed in the region until at least 10,000 years ago, it is quite possible that the Mapinguari legend has its origin in genuine encounters with these large, hairy, huge-clawed creatures. As with Bigfoot, Yeti and the Loch Ness Monster, however, reported sightings have continued, and some maintain they are evidence that some species of giant sloth live on in the region. As one American anthropologist and collector of Mapinguari tales put it, 'There's still an awful lot of room out there for a large sloth to be roaming around.'

In 1901, however, an expedition failed to find one. On 12 July of that year, the *Daily Express* reported:

The expedition to Patagonia in search of the giant sloth has returned without having discovered it. Hesketh Prichard, chief of the expedition, during nearly a year spent on the eastern side of the Cordilleras, found some remains of the giant sloth bearing an extraordinary appearance of freshness. He discovered a species of puma new to science, and a new lake in which were many icebergs. Large zoological, ornithological, geological and botanical collections were brought back.

In other words, a few animals, birds, rocks and plants – but no Mapinguari.

In 1994, another expedition to find a Mapinguari was reported in the *New York Times*. Led by Dr David C. Oren, an American ornithologist working for the Brazilian Government, the team headed for the depths of the western Brazilian rainforest in search of a living animal the paper described as 'a human-size ground sloth belonging to a family thought by palaeontologists to be long extinct'. The report quoted locals describing the creature as 'terrifying and dangerous, physically powerful and equipped with some kind of chemical defense capable of paralysing opponents', and said that Dr Oren had conducted more than a hundred interviews over nine years 'with Indians and rubber tappers who told of having had contacts with the creature'. Their descriptions seemed to tally with other accounts of the Mapinguari. 'When I began hearing accounts of a creature with shaggy red hair, backward-turned feet and a monkeylike face,' Dr Oren said, 'I realized that witnesses might have encountered a ground sloth, closely related to extinct giant sloths known only from their fossils.'

From these descriptions, they hoped to encounter a creature some six feet tall, weighing around 500 lbs, with jaws and feet that could tear palm trees apart. It also had a thunderous voice that sounded almost human, and was said to live on palm hearts and other vegetarian delicacies. Sadly, they were unable to find it. In fact, the only physical evidence was a set of footprints that seemed to point backwards.

Quite apart from such mythical creatures as the Mapinguari, the culture of the indigenous peoples of Central and South America contains a number of myths and old tales about sloths. Perhaps the most typical (and slothful) of these is a story from the Karaja tribe of the Brazilian Amazon, which explains why sloths or '*ay-ees*' as they called them from their sigh, are the only arboreal creatures which do not build nests or some sort of homes in the Cecropia trees known as 'monkey-paw' from the shape of their leaves. It goes something like this:

Tomorrow the Sloths Will Build Nests

The night was stormy and the wind lashed at the mothers and baby *ay-ees*. They shivered as the rain soaked them and they lay unprotected on the branches of the monkey-paw trees, and they sighed '*Ay-ee, ay-ee*'. Seeing and hearing this, a father *ay-ee* made a firm decision: 'Tomorrow, we build nests!'

'Yes,' the other fathers agreed. 'Tomorrow for certain.'

However, the next morning was sunny and the father *ay-ees* basked in the warmth of the sun for some time to dry off. Then they settled down to eat their breakfast, slowly, and for a long while.

After breakfast, it was time for a nap to gather strength; a long, long nap, which lasted a long, long while, after which they all felt so well that nobody thought of building nests. At least, not until the next downpour, when another storm raged and the wind and rain lashed at the trees where the mothers and baby *ay-ees* were trying to sleep.

And what did the father *ay-ees* say then?

'Tomorrow, we will build nests!'

Another Brazilian folk tale, however, features a sloth that has indeed made its home in a tree, though one might guess from the ending that it was the last one to do so:

The Sly Sloth

Once there was a sloth that lived deep in the rainforest, and even for a sloth he was very slothful. In fact he was the laziest sloth in the forest and was so lazy that when he ran out of food, he could not be bothered to find more. Instead, he asked his four closest neighbours if he could share some of their food, promising to repay them by inviting them to a great party at his house the next day.

First, he borrowed food from a mouse as it walked past his house. Then he borrowed food from a snake; then a wild boar; and finally a jaguar. Without even leaving his home, he had amassed a real feast, half of which he ate that evening, then the other half in the morning after a good sleep.

The sloth then wondered what to do, as the other animals would all be coming round, expecting to be repaid with a great party. Then he hatched a plan, which began with him wrapping a cloth round his head and going to bed, pretending to be ill.

The mouse then arrived at the sloth's house and felt sorry for him. Before he could mention the food, however, the snake knocked on the door.

'Quick,' said the sloth to the mouse. 'Hide under the bed or the snake will eat you.' And the mouse darted under the bed.

Then the snake came in and also felt sorry for the sloth, but before he could ask for his food back, the wild boar knocked on the door.

'Quick,' said the sloth to the snake. 'Hide under the bed or the wild boar will eat you.' And the snake slithered under the bed.

Then the wild boar came in, but before he could say anything, the jaguar knocked at the door.

'Quick,' said the sloth to the wild boar. 'Hide under the bed or the jaguar will eat you.' And the wild boar squeezed under the bed.

Then the jaguar came in and noticed a great commotion going on under the sloth's bed. For the other animals had all become aware of each other, and the snake ate the mouse, and the wild boar ate the snake, whereupon the jaguar pounced and ate the wild boar.

While all this was going on, the sloth was able to sneak out of the house and hide from the jaguar. And he is still hiding from jaguars to this very day.

The next folk tale comes from the Arawak peoples of South America and the Caribbean. It has a sad ending for the sloth and seems to view people in a particularly bad light too:

The Woman in Love with a Sloth

A woman had a Sloth for a sweetheart. Whenever she went into the field or into the bush she used to bring food and drink for him. She would call 'Hau! Hau!' and the Sloth would clamber down the tree to join her and they hugged and caressed like lovers.

Seeing the woman continually taking food and drink out of the house, other people began to talk, wondering what she did with it or who she was taking it to.

One young man kept watch on her, followed her, and saw her call her Sloth lover and caress him. This time, however, instead of returning her caresses, the Sloth scratched her, and pulled her hair.

'Are you jealous of me, or worried by something?' the woman asked the Sloth, but the Sloth could not reply. In fact, the Sloth was very worried, as well as jealous, because he could see the young man watching from behind a tree everything they were doing. The woman did not know this, and returned to her home. As soon as she was gone, the man came out of his hiding place, and killed the Sloth.

When the woman returned next day, she saw the animal lying dead, and fell into a great grief and wept bitter tears, saying, 'What has killed you, my darling?'

But the young man, who had been following her, then came close to her, and consoled her. 'Don't be foolish,' he remarked. 'A fast fellow is preferable to a slow Sloth. Take *me* for your sweetheart.' And she did.

I cannot help feeling that they were not destined to live happily ever after: the fellow seems a total slothicidal brute and I really cannot understand what she saw in him.

A deeper analysis of the role of sloths in South American legend may be found in the work of the French anthropologist and ethnologist Claude Lévi-Strauss. His book *The Jealous Potter* (1985) is a characteristically profound and imaginative investigation of the ways in which folk tales ascribe certain personality attributes to the practitioners of certain trades, and the similar tendencies of story writers to impute typical anthropomorphic tendencies to certain animals. Chapter 7 of the book, which is entitled 'The Sloth as a Cosmological Symbol', deals not only with the sloth itself but, somewhat surprisingly perhaps, with the excrement of the sloth. He begins with a tale from the Tacana people of Bolivia which explains the origins of present-day humans. Similar myths also occur in other tales from neighbouring regions and the following summary is typical:

The Power of Poo

Long, long ago, humans knew nothing of fire and lived by breathing the wind. One day a man brought home a sloth for his two children and the animal lived at the top of a tree where it ate the leaves.

The children, however, annoyed the sloth by not letting it come down the tree to defecate on the ground. Even when the sloth threatened to kill them, the children kept harassing him, so he dropped to the ground and relieved himself.

When his poo hit the earth, the ground began to smoke. Soon flames appeared, the fire spread and large cracks appeared which swallowed up all the human beings, carrying them into the underworld. Only when the fire had died down and the earth had cooled did another race of humans emerge from the underworld by climbing up a series of bamboo sticks.

Lévi-Strauss also tells us that in one version of the same story, the sloth is said to have created flatulence (despite everything we said at the end of the previous chapter about sloths not farting), though whether this is human flatulence, sloth flatulence, or both, is not made clear. He also tells us of an ancient belief of the Yagua people of Colombia and northern Peru that two sloths with human heads hold up the world at both ends, preserving its balance in the skies. The Machiguenga people of southern Peru, however, gave the name 'The Sloth' to the Large Magellanic Cloud, which is a satellite galaxy orbiting the Milky Way. They believed this was the only illumination of the sky before the moon came into existence.

Quite what the sloth did to become associated with such myths is not totally clear, but Lévi-Strauss was convinced that the link between sloth and myth was there:

From the Tacana in eastern Bolivia all the way to the Kalina in Guiana, passing, on the way, through a whole series of other peoples, the Sloth thus stands as a cosmological symbol. It is particularly clear in the Tacana myths, but in others, too, that this role is linked to the animal's habits, especially to those concerned with the functions of elimination.

Hunting the sloth. An illustration from *Chatterbox* (1906) that accompanied a story from Charles Waterton's *Wanderings in South America* (1828).

Chapter 11

SLOTHS EATING, AND EATING SLOTHS

'I would come back as a sloth. Hanging from a tree, chewing leaves sounds great.'

David Attenborough, interviewed
in *Time Magazine* (2011)[*]

WARNING: The second half of this chapter may not be suitable for any person of heightened sensitivity, but no book on sloths can be complete without including its subject. Don't worry: we shall warn you again when we get to the gruesome bit. And the first half is OK for anyone.

Gastronomy 1: What Do Sloths Eat?

Three-toed sloths are very fussy eaters. Their favourite food is undoubtedly the tender, young leaves of the Cecropia tree, which is

---

[*] In reply to a question on which animal he would like to come back as if he were reincarnated. Sadly, when asked about this in 2018, Sir David said he considered it a frivolous answer to a frivolous question.

why the best place to look for them is in the upper branches of that tree. Indeed, it was for a long time thought that Cecropia leaves were the only thing the Bradypus would eat, but recent surveys show that they eat other leaves as well – if not quite so enthusiastically. Indeed, while individual sloths exhibit a wide range of personal tastes in the leaves and shoots they will eat, Cecropia leaves seem to be the only variety that satisfies all individuals.

Cecropia trees, however, are indigenous only to Central and South America, which explains why the sloths in most zoos in other regions of the world are of the two-toed variety. Early attempts to keep three-toed sloths in captivity are known to have frequently ended in the poor creatures starving to death because Cecropia leaves were unavailable.

Two-toed sloths are much less fussy and have been known to eat a wide variety of foods as well as Cecropia leaves, including several vegetables, nuts, fruits, berries, caterpillars, birds' eggs, and even young birds, mice and rats.

Laurie J. Gage's 2008 book *Hand-Rearing Wild and Domestic Mammals* includes a chapter on sloths written by Judy Avey-Arroyo in which she gives detailed information on feeding young sloths in her renowned Aviarios del Caribe Sloth Rescue and Rehabilitation Centre in Costa Rica. Her experiences include successfully bringing up baby orphan sloths on goat's milk (cow's milk is a definite no-no: it gives the baby sloths fatal diarrhoea) and weaning them around the age of four months on to leaves, in the case of three-toed sloths, or vegetables for two-toed. Their success rate for the two-toed species far exceeds the rate for three-toed varieties.

Cincinnati Zoo reaches much the same conclusions, reporting that 'Mr. Two-Toes in captivity – where he is much more likely to survive than his three-toed counterpart – dines on special dry food made for leaf-eaters, along with a wide assortment of fruits and vegetables, with grapes as his particular passion.' London Zoo also feeds its sloths on special vitamin-enhanced food pellets, specially designed for rainforest inhabitants.

There is one important aspect in which sloth diets differ from those of other mammals, however, and that is the way they react to changes in temperature. A paper published in 2015 with the title 'Sloths Like It Hot: Ambient Temperature Modulates Food Intake in the Brown-throated Sloth (*Bradypus variegatus*)', in the open access journal *PeerJ* (by Becky Cliffe and others), reported the results of a detailed study measuring the amounts of food eaten every day over a five-month period by three captive animals at the Sloth Sanctuary in Costa Rica. They were all given the same amount of Cecropia leaves twice every day and leftovers were weighed to determine the amount eaten. Results were then correlated with the temperature and showed that the warmer it became the more the sloths ate.

This is the opposite from most mammals, which eat more in colder weather in order to generate the energy needed to keep warm, but the result makes sense for sloths for two reasons. First, unlike most mammals, their body temperature varies with the weather, so they do not need as much of the energy provided by food in order to maintain a constant temperature, and secondly, their stomachs are full most of the time anyway as they may take as long as fifty days to digest the leaves they munch. Hot weather leads to the food in their stomachs fermenting more quickly and moving into their intestines. This leaves room for more food intake into their stomachs, so could explain why they eat more in hot weather.

We move on to the part responsible for the warning at the start of this chapter. Sensitive readers may be well advised to skip this section.

Gastronomy 2: Eating Sloths

Alan Davidson's magnificent *Oxford Companion to Food* includes no mention of sloths. If you try to look them up, you will find that he leaps directly from SLOE (which we learn makes a good imitation of the Japanese dried and salted plums known as *Umeboshi*) to SLOVENIA (which has a fine tradition for types of porridge). You will also find no entry for Bradypus (the book skips from BRACKET FUNGI to BRAINS) and there is no Choloepus between CHOLESTEROL and CHOP.

Yet there are areas of the rainforest in Central and South America where sloths account for between a quarter and a half of the biomass of the large mammal population. We have also seen that the most likely reason for the extinction of giant sloths around 10,000 years ago was that they were hunted by human beings, probably for food. In view of all this, it would seem likely that the occasional sloth might be seen as a tempting meal for modern carnivorous humans, and that does indeed seem to be the case.

Fortunately, hunting sloths is against the law in most, if not all, of the countries where they are commonly found. Costa Rica has recently even made it illegal to cuddle or hold a sloth or to pose with it for a selfie. Sloths are very sensitive animals and do not like being picked up by strangers. It has also been reported that several indigenous tribes in Brazil have taboos against hunting or eating sloths. This is all good news for the sloths, but despite this, there are

> As the first shock of the change of light passed, I saw about me more distinctly. The little sloth-like creature was standing and staring at me.
>
> H.G. Wells, *The Island of Dr Moreau* (1896)

some accounts of them ending up on the dinner table.

The most exhaustive account of sloth-eating is to be found in the 2006 PhD thesis of anthropologist Jeremy M. Koster at Pennsylvania State University entitled 'Hunting and Subsistence among the Mayangna and Miskito of Nicaragua's Bosawas Biosphere Reserve'. Combining his own research with that of others, he produces a remarkable table of the meat-eating habits of eleven tropical societies. Having collected data on the sloth-eating practices of all but one of these eleven, he reveals that sloths are eaten by virtually all the members of three groups, by some members of another two, but that five tribes never eat sloth. He also tells us that 'I encountered no societies that eat three-toed sloths but not two-toed sloths', and quotes another earlier researcher who found that sloth meat, whether two-toed or three-toed, 'leaves something to be desired'. Nevertheless, Koster tells us, with perhaps dubious precision, that the average Neotropical tribesman eats 0.64 of a sloth per year.

More detailed information on the preparation and taste of sloth meat was provided in an interview given in 2012 by the American missionary and linguist Daniel Everett, who spent more than seven years living with the Pirahã tribe in Brazil, studying their customs and language. Their basic method of cooking sloth, he said, was to singe off the hair, clean its guts out, then roast chunks of muscle over an open fire before tearing off pieces and eating with the hands.

Everett appears to have been unimpressed with the taste, and is quoted as suggesting that 'spices may do wonders for the palatability of sloth meat'. His own recommendation is to tenderize the meat in a pressure cooker for forty minutes, with coriander, garlic, salt and chili sauce, and then to eat it with tacos.

Another American, the composer Aaron Paul Low, helped catch and eat a sloth on a trip to Peru in 2012. He described it as 'one of the few absolutely disgusting animals we ate. It was really, really tough, and there really wasn't that much meat.' He also described the taste of a turtle as 'truly revolting' and not a patch on that of a crocodile, which was 'somewhere between chicken and shrimp'. This is very different from the view of British naturalist and explorer Alfred Russel Wallace (1823–1913) who wrote, in his *Travels on the Amazon* published in 1853: 'Our men had caught a sloth in the morning, as it was swimming across the river... The Indians stewed it for their dinner, and as they consider the meat a great delicacy, I tasted it, and found it tender and very palatable.'

Peter Lund Simmonds, in his 1885 book *The Animal Food Resources of Different Nations: with Mention of Some of the Special Dainties of Various People Derived from the Animal Kingdom*, referred to this passage by Wallace, but went slightly further, suggesting that the Indians 'hunt the animal for the purposes of food', rather than just happening to eat what they caught.

Wallace, of course, is now best known as joint founder of the Theory of Evolution and the man who suggested to Charles Darwin that he use the phrase 'survival of the fittest', which had first been used by the anthropologist, biologist and social theorist Herbert Spencer.

The subject of hunting sloths for food brings back a memory of Charles Waterton, who wrote of the sloth, in 1825: 'He is a scarce and solitary animal, and being good food is never allowed to escape destruction by the Indians [who] use arrows tipped with the *wourali* poison in its destruction. The flesh is so much relished by them that they are in continual pursuit of it.' Advising against pointing a gun at a sloth, or using a poisoned arrow, he also wrote: 'His looks, his gestures and his cries all conspire to entreat you to take pity on him. These are the only weapons of defence which Nature hath given him... It is said his piteous moans make the tiger relent and turn out of the way.'

Humans, however, have clearly not all been moved by such pity, even if not motivated by the quest for food. In 1896 the *New York Times* ran a piece called 'Hunting the Sloth', written by someone with the inappropriately benevolent pseudonym 'St Nicholas':

> The tamest hunting in the world is sloth hunting, in comparison with which the pursuit of orchids is quite exciting, and turtle-catching is wild and dangerous sport. But I have done my turn at it, nevertheless. Once, on the mighty Esquibo River, in British Guiana, I took a native companion, a gun, an axe, and a leaky canoe, and set forth to round up a lot of chestnut-headed sloths.

We paddled about thirty miles that day, and picked eight sloths. They were found by paddling along the shore, and watching the tree-tops for things that looked like big grey spiders. Sometimes sloths 'spread-eagled' on the outer branches of a tree; others would be hanging upside down, but always eating. They eat so slowly that before one meal is over, it is time for the next. Usually the gun would bring them down, but sometimes it was not necessary. Two were taken alive by Paulie, who climbed up and plucked them like so much fruit, and twice we had to cut down trees.

I should also mention one fictional account of eating a sloth, from the 1904 novel *Green Mansions: A Romance of the Tropical Forest* by William Henry Hudson (1841–1922), which is set in a South American rainforest. Having spent most of his life in Argentina, Hudson's culinary description in this passage of what to do with sloth leftovers is presumably based on experience:

> In the evening of that day, after completing my preparations, I supped on the remaining portions of the sloth, not suitable for preservation, roasting bits of fat on the coals and boiling the head and bones into a broth; and after swallowing the liquid I crunched the bones and sucked the marrow, feeding like some hungry carnivorous animal.

Last (but not least) of all, some account should be presented of an alleged meal of sloth meat that was probably not a sloth at all.

In 2016, a remarkable paper appeared in the online science journal *PLOS One*. Entitled 'Was Frozen Mammoth or Giant

Ground Sloth Served for Dinner at the Explorers Club?' and written by a team led by Jessica Glass of Yale University, it told of investigations into a notorious meal served at the Club's annual dinner in New York in 1951. The menu, which purported to include frozen mammoth from Alaska that was over 200,000 years old, quickly became legendary, but doubts were soon raised over the true contents of the meal.

Although one of the diners, in an interview a few days after the meal, clearly stated that they had eaten mammoth, several others explicitly reported that it had been giant sloth. Since other Explorers Club annual dinners, both before and after 1951, are reliably reported to have served fried tarantulas, goat eyeballs and bison and horse that had been preserved in permafrost, extinct mammoth or sloth was not out of the question.

A sample of the meat had, however, been sent to Paul Griswold Howes, a member of the Explorers Club, who had been unable to attend the dinner. Howes was also a museum curator who promptly displayed the meat in preservative at the Bruce Museum in Greenwich, Connecticut. The provenance of the meat was certified by Commander Wendell Phillips Dodge, who was Chairman of the Explorers Club Annual Dinner Committee. But the specimen was clearly labelled 'Megatherium' not 'Mammoth'. In other words, it was a giant sloth after all.

This only added to the speculation, as there had never been any evidence that giant sloths had penetrated as far north as Alaska, so quite apart from the question of the true nature of the animal it had come from, there were doubts about whether it was from a body preserved in Alaskan frost at all.

What led to the 2016 paper was that DNA analysis had by then reached a stage that offered the possibility of extracting DNA from the meat sample and comparing it with that of extinct giant sloths and mammoths. So that's what they did.

And the results reported in *PLOS One* showed that the meat was almost certainly that of a green sea turtle, which tied in very well with the fact that turtle soup had also been served at the notorious 1951 dinner.

Anyone wishing to continue believing that extinct giant sloth was eaten in New York in 1951 may, of course, cling to the belief that Commander Dodge simply sent the wrong piece of meat to the museum curator, but after all the tasteless material in the chapter, it's good to have ended on an alleged sloth meal that probably wasn't sloth after all.

Chapter 12

SLOTH CONSERVATION

We find the slothful watch but weak.
William Shakespeare, *Henry VI, Part 1* (c. 1592)

For millions of years, as we outlined in the early chapters of this book, giant ground sloths ran amok around South America while the much smaller tree-sloths hung peacefully in trees. Then the giant sloths died out and for thousands of years, tree-sloths and humans co-existed, usually in blissful ignorance of each other, or at least in reasonably blissful tolerance, with their respective ways of life having little contact with one another. In the last couple of decades, however, human activity has been having an increasing effect on sloth-kind, encroaching on their territory and in some cases threatening their very existence.

Fortunately, we have also begun to notice what extraordinary creatures they are, and increasing conservation efforts are being made: particularly in the case of the maned sloth of south-eastern Brazil, which is listed by the IUCN (International Union for the Conservation of Nature) as 'vulnerable', but also for the pygmy sloth of Panama, which has been listed as 'critically endangered' since 2006. To trace the development of conservation efforts, both on an individual and countrywide basis, we must go back to the early 1990s, before which sloths and humans had very little to do with one another. Sloths mostly hid away at the top of trees in the dense areas of the rainforests of Central and South America unseen by the vast majority of people and generally ignored by those who did catch sight of them. Even now in most parts of Costa Rica, where the sloth is seen as a great boost to tourism, most people have grown up to think of the sloth as a type of vermin.

That all began to change in 1992 when three girls brought a baby sloth to the Aviarios del Caribe in Costa Rica, a biological reserve designed mainly for bird-watchers. The girls knew it was run by animal-lovers and it was the best place they could think of to bring a young sloth whose mother had been run over by a car. The timing was fortuitous. The previous year, a huge earthquake had struck the area around Limon where the Aviarios was situated, which changed the course of a river, down which bird-spotting tours had been run. So the bird tours ceased and the reserve's founders and owners, Judy Avey-Arroyo and Luis Arroyo, had time to shift their attention to their new orphan guest, whom they called Buttercup. Their attempts to contact zoos and other wildlife centres, however, led them quickly to the realization that nobody knew how to bring up a sloth, so they drew up their own rules by watching how wild sloths did it. Their success in doing so is marked by the fact that Buttercup is still with them more than twenty-five years later.

Once word got around about the novel activity of Aviarios del Caribe, more people begin to bring injured or orphaned sloths to them to look after, and they changed their name and started conducting guided tours. In 1997, the Sloth Sanctuary of Costa Rica was officially recognized as a rescue centre and quickly became one of the country's best-known and most popular tourist attractions.

This brought increasing numbers of injured and orphaned sloths as well as tourists to the Sanctuary. (Injured sloths, incidentally, are sent to an area known as the 'slothpital'.) Over the past twenty-five years, over 750 sloths have been brought there, which in 2016 led to accusations of overcrowding, insufficient care for sick sloths, and suggestions that too few sloths were being returned to the wild. And meanwhile, similar questions were beginning to be asked about the many other sloth sanctuaries that were beginning to spring up in Costa Rica, and other countries in Central and South America. According to one recent estimate, there are now close to a hundred of these, but they may range from true sanctuaries to little more than zoos, motivated more by a desire to profit from tourism than genuine concern for sloths.

Such accusations, however, have drawn attention to how little we really know about sloths. How much space do they need? What can they be safely given to eat? Under what circumstances, if any, can a baby sloth brought up in captivity be safely returned to the wild? An earthquake and an orphaned sloth may have sown the first seeds of sloth tourism, but they have also coincided with a vast increase in research and conservation efforts. The attempts at conservation have arrived not a moment too soon.

While a few people like the Arroyos were beginning to show an interest in sloths, the habitat of these animals in many places was being severely encroached upon by humans. The Amazon rainforest is estimated to be losing 20,000 square miles in area every year through deforestation, large-scale cattle operations and other developments designed to support mankind's insatiable demands for certain foods. Not only does this restrict the area in which sloths may live, but it also reduces their food supply. With three-toed sloths in particular largely reliant on the younger leaves of mangrove and Cecropia trees, large-scale deforestation can be a disaster for them.

As if deforestation were not bad enough, it is not the only way human encroachment is damaging to sloths. Not long ago, the main threats to a sloth's peaceful existence came from ocelots and other wildcats on their weekly trips to ground level, and from harpy eagles and the occasional owl when they were high in trees. Now cars and electricity power lines are responsible for increasing mortality rates, while sloth tourism has also brought undesirable effects with an increasing number of cases of sloth capturing to supply the illegal pet trade. Even well-meaning tourists can have a bad effect: several sloth conservation centres report an increasing number of sloths being picked up in the road and brought in by people thinking that the poor animals must be lost while they were probably just moving slowly from one tree to another for dinner. Helping them across the road is one thing, but carrying them to a sanctuary miles away is not helpful.

Furthermore, in 2017, an analysis by World Animal Protection of sloth selfies on Instagram showed that such pictures, typically arranged by unofficial tour brokers, has increased by nearly 300 per cent since 2014, and investigations showed that many such photos were the result of sloths being illegally captured in rainforests and held captive where tourists would pay for pictures. Studies have shown that sloths become alarmed and anxious when picked up by people they do not know and there are campaigns to ban, or at least heavily discourage selfies with sloths. In this respect, the poor creatures have become too popular for their own good.

Other human activities, however, are having far more serious effects on entire sloth populations. Increasing urbanization has led to ever increasing demands on land for homes and cattle-farming. In many areas, what used to be continuous forest has turned into isolated segments, bringing danger to sloths as they attempt to cross from one to another. Quite apart from the destruction of their former habitat, they face the perils of road collisions, attacks by dogs, human cruelty and electrocution from power lines.

Until recently, sloths have been remarkably successful in propagating their species. In some rainforest areas of Central and South America, surveys have estimated that they constitute up to 70 per cent of the total arboreal mammalian biomass. Even if we include the ground-based mammals, sloths may account for between one-third and one-half of their total mass. Yet rapid changes to their habitat can be a real threat, and this is being met by the increasing effort being given to conservation projects, particularly for the threatened species of maned and pygmy sloths.

Quite apart from the danger to sloths themselves, their bodies provide a home for dozens of species of moth and algae that are found nowhere else, and we have no real idea what part this ecosystem plays in the wider ecosystem of the rainforest. The sloth's extraordinary capacity to recover quickly from injury has also been cited as a potential source of medical advancement, though nobody yet has the faintest idea how they do it. For those who are unconvinced of the wonder of sloths, however, potential medical benefits may be a convincing reason for making conservation efforts.

In 2002, Brazil launched a project known as ARPA (Amazon Region Protected Areas), with the goal of turning 150 million acres

> # Awake, awake, English nobility! Let not sloth dim your horrors new-begot.
>
> William Shakespeare, *Henry VI, Part 1* (c. 1592)

of the Brazilian Amazon rainforest into a combination of sustainable use and strict protected areas. This is said to be the largest conservation project in the world and its latest development is the ARPA for Life initiative, supported by the World Wide Fund for Nature (WWF), which has created a $215-million fund to finance its activities. Naturally, this is not just for sloths, but that is the animal chosen to symbolize the project and lead the fund-raising campaign.

On a much smaller scale, but far more urgent in view of the species they protect, the Zoological Society of London's EDGE of Existence programme (a splendid acronym standing for 'Evolutionarily Distinct and Globally Endangered') is making great efforts on behalf of the critically endangered pygmy sloth, which is found only on the island of Escudo de Veraguas, off the east coast of Panama. The island itself, which is only 4.3 square kilometres in area, began to drift away from the Central American mainland around 10,000 years ago and is now a four-hour boat ride away. Over that time, the three-toed *Bradypus pygmaeus* has evolved to be about 40 per cent smaller than other sloths. It has been recognized as a distinct species only since 2001 and there is a still a great deal of vital information to find out about them in order to optimize conservation efforts.

The first question to ask is how many pygmy sloths there are on the island. A survey published in 2012 reported that only seventy-nine had been found, but the best-informed estimates now suggest that the correct figure is around 500, the most optimistic that it could be as high as 3,000. In 2015, leading sloth researcher Bryson Voirin was quoted as saying that the 'actual population size is most likely somewhere between... perhaps 500 to 1,500 individuals'.

The trouble is that pygmy sloths were initially seen only in the island's mangrove regions, which made good sense, as mangrove leaves seemed to be their favourite food. But mangroves cover only 3 per cent of the island, which also has a vast amount of unsurveyed dense forest area.

Voirin's 2015 estimate was based on the results of a three-year survey that had involved placing radio-collars on ten sloths and tracking them at three-to-six-month intervals. He found that only three had remained in the mangroves and that the others had wandered off into the dense forest. What they ate there is still a mystery, but these findings raise the possibility of there being far more pygmy sloths on the island, hiding away in the forest, than had previously been imagined. In October 2014, a preserve of pygmy sloths was officially confirmed deep in the forests and researchers said they were 'slowly establishing trust' with the indigenous people to help create an effective conservation programme. More accurate estimates of the total pygmy-sloth population are now being sought by the EDGE researchers trialling GPS devices on the sloths, including both backpacks and collars, with accelerometers to track their location and movement, but sloth-counting and tracking is only a small part of their efforts.

In 2009, a vast area around Escudo de Veraguas was designated a Protected Landscape by the Panamanian Government, of which the island itself is the only land area. There are no permanent residents on the island, but seasonal visitors such as fishermen, lobster divers, tourists and members of the local Ngäbe-Buglé people harvest timber from the mangrove trees to maintain wooden houses on the island.

There has also been talk of efforts being made to exploit the tourism potential by building an ecolodge and marina, and even rumours of a casino and banking centre, to turn Escudo into an offshore tax haven. With no staff appointed by the Panama Government specifically to enforce protection to the island, and disagreements between regional and national politicians and local administrators, further complications are always liable to hit measures designed to protect sloths.

Nisha Owen, who heads the ZLS's EDGE pygmy sloth project on Escudo, lists among their aims, alongside population and ecological surveys, building conservation capacity, and detection of threats such as hunting and mangrove extraction, mitigating such threats 'by design of outreach programmes to increase local awareness and enhance public support – also in mainland communities'. This includes an educational programme in elementary schools 'to raise awareness of the importance of the pygmy sloth and Escudo in terms of ecosystem services, such as the provision and maintenance of natural resources, and as a part of the Ngäbe-Buglé heritage; in doing so, we aim to engender support for conservation'.

Meanwhile her colleague Dr Diorene Smith, who leads the conservation and outreach working on the ground in Panama, has helped to establish a 'Committee for the Protection of the Pygmy Sloth' from what they call 'anthropogenic disturbance', which reviews, and makes recommendations on permissions for, all relevant human activities in the area, from scientific research to tourism. In short, they hope to educate the local community into realizing the uniqueness of these sloths and the potential benefits they can bring to the region, and into taking a proper pride in them. With EDGE researchers providing scientific support, the local community will then be eager to do the necessary work to ensure that sloths and humans can live in harmony, including proper arrangements for the collection and recycling of increasing amounts of rubbish being left on the island by visitors.

With so few pygmy sloths on Escudo, the matter is clearly urgent, but EDGE have only been working there since 2012, so a recent increase in sightings of other sloths with babies suggests that they are making good progress and the sloth population has at least stabilised.

Finally, before leaving the subject of sloth conservation I should mention a little more about SloCo (slothconservation.com), the registered non-profit organization founded by Becky Cliffe and dedicated to 'bringing together all people, partners, and institutions working globally with sloths to achieve lasting conservation solutions'. The list of conservation strategies on which SloCo is working include:

- Education programmes in local communities;
- Campaigning for underground power lines;
- Planting forest corridors to connect fragmented habitats;

- Construction of wildlife bridges across major roads;
- Tree surveys to ensure that the species needed to support sloths are plentiful enough;
- International education programmes to reduce poaching for the sloth pet trade and tourist photo opportunities.

With programmes such as those of the WWF in Brazil, EDGE in Escudo and SloCo in Costa Rica all sharing ideas with each other, and even the local Brazilians, Panamanians and Costa Ricans beginning to appreciate the wonder of sloths, there are grounds for optimism, though much work remains to be done.

And while we're on the general topic of the potential benefits of sloth conservation, we should perhaps end by mentioning a paper, by Sarah Higginbotham and others, published in 2014 in the online science journal *PLOS One*, entitled 'Sloth Hair as a Novel Source of Fungi with Potent Anti-Parasitic, Anti-Cancer and Anti-Bacterial Bioactivity'. It was already known that several species of moth were only found living in sloth fur, but this research isolated over eighty fungi in the algae growing there, many of which were previously unknown to science.

Twenty of these were found to be effective against strains of parasite that cause malaria, have a wide range of antibacterial properties, and have potential in the treatment of human breast cancer. The researchers conclude by saying: 'Thus our work suggests that fruitful exploration of the sloth microbiota is warranted for potential applications in drug discovery.'

Chapter 13

SLOTHS IN CULTURE

**Sloths move at the speed of congressional debate
but with greater deliberation and less noise.**

P.J. O'Rourke, *All the Trouble in the World* (1994)

Peaceful, unpretentious and shy, sloths have lived their lives at the top of Cecropia trees for over 30 million years without making much of an impact on our culture until recently. Indeed, outside the world of palaeontologists trying to find out about giant sloths, the animals have been practically invisible. The British Library catalogue contains references to sixty-four books with the word 'sloth' in their title and the majority of those are religious tracts discussing the Deadly Sin that shares the animal's name.

Until the 1990s, you could count the number of true sloth books on the fingers of one hand. I would love to have said 'on the claws of one three-toed sloth hand', but unfortunately there were four such books. First, in 1953, came a children's book called *Julie's Secret Sloth* by Jacqueline Jackson. This was followed by *A Sloth in the Family* (1966), Hermann Tirler's delightful account of his adoption in South America of a sloth which he named Nepomuk because of its supposed resemblance to a statue of St John of Nepomuk in Prague. In the book, Tirler refers to 'something in his face that reminded me of the statue of St Nepomuk in Prague', which was a happy mistake: 'Nepomuk' is a far better name for a sloth than 'John'. Despite that, neither Nepomuk nor Julie's secret sloth seemed to capture the public imagination enough to set off a wave of sloth books.

The next title was *Function and Form in the Sloth* (1971) by M. Goffart of Liège University in Belgium, which was for more than four decades the standard academic treatise on sloths, but could hardly be described as popular reading. This was followed in 1976 by a collection of photographs by John Hoke (1925–2011) called *Discovering the World of the Three-Toed Sloth*, but again sloths were slow to take off into the realm of best-sellers.

The cigar-chomping, pith-helmeted American adventurer Hoke, incidentally, was said to have 'become enamoured of sloths' in the 1950s in Suriname after one bit him. He brought a sloth back with him to Washington, named it Lady III and donated it to the National Zoo. He also wrote books on snakes and turtles for the children's publishing house run by his mother and, perhaps most remarkable of all, he built an electricity generator run by squirrels on treadmills.

Hoke worked for the US International Development Agency until 1962, when he was sacked for trying to spend a vast amount of money on the development of a solar-powered boat. When called to testify before the Agency, he mentioned his writings while in Suriname:

> As a recreational endeavour, I spent a lot of time prowling around in the bush and I studied an animal. I won't bore you with the details of the type of animal, but I did a study on this

animal because it is not well known, and wrote a definitive piece on this animal.

Later in his testimony, Hoke's questioners renewed their interest in this animal and the following exchange took place which confirms the lack of general knowledge about sloths at the time:

Mr HOKE: I discovered an animal there that I could find no writing on. This was the three-toed South American sloth. It is a small animal that people pay no attention to, and for good reason. It does not do anything.

Mr HARDY: What is a sloth?

Mr HOKE: It is a slow-moving animal that has no economic importance.

Mr HARDY: This does not belong to the frog family, does it?

Mr HOKE: No. It is a mammal. It lives in the trees and it is a rather endearing creature. Someone gave one to us as a pet. People brought animals to me because they knew I was interested in wildlife and I was interested in this animal but I could find nothing written on it, and all my efforts to find something written on it resulted in the fact that I knew as much as anyone else on it, so I wrote about it.

Mr HARDY: I would like a chance to read that. That sounds interesting.

The Hon. Porter Hardy Jr, who was the chairman of the House of Representatives Foreign Operations and Monetary Affairs

Subcommittee, later indicated that he would like to see a photograph of a sloth, but Hoke did not have one with him at the time. Hardy later made a probably cynical comment that Hoke might be able to use sloths to power an electricity generator, but that was the last time sloths were mentioned in the long deposition in August 1962.

It took until 1990 for things in the world of sloth books to begin to change, but matters then began to move with greater rapidity at least in the realm of children's books. *So Slow Sloth* (1990) by Charles Fuge and *Sloth's Shoes* (1997) by Jeanne Willis and Tony Ross took up the baton that Jacqueline Jackson had set off with in 1953, and Dick King-Smith ran with *The Great Sloth Race* in 2001, but the book that really made publishers realize that children could love sloths was Eric Carle's *'Slowly, Slowly Slowly,' Said the Sloth* in 2002.

With this acclaimed author of *The Very Hungry Caterpillar* now on board, sloth books for children suddenly gained great popularity. In 2003, Rose Impey introduced us to *Sleepy Sammy: The Sleepiest Sloth in the World*, while in the same year Andrew Murray brought out *The Very Sleepy Sloth*.

Since then, we have seen *Diego Saves the Sloth!* (2008, by Alexis Romay), *The Sloth Who Could Not Sleep* (2011, by Samantha Brenton) and *Sloth Slept On* (2014, by Frann Preston-Gannon).

Not to be left behind, the non-fictional sloths struck back: a book by Susan B. Neuman called *Swing, Sloth!*, subtitled *Explore the Rain Forest*, was published by National Geographic Kids in 2017; and Lucy Cooke's best-selling books of pictures and tales from the Aviarios Sloth Sanctuary in Costa Rica, *A Little Book of Sloth* (2013) which was published in the UK as *The Power of Sloth* (2014).

Sloths for adults still lagged well behind, though Gillian Bridge in 2007 celebrated the joy of sloths in her own way by bringing out the self-help book *Discover Your Inner Sloth*. Before that, references to sloths in literature were rare. The plays of Shakespeare contain six mentions of sloth, but they all pre-date the name being given to the animal and all refer to the sin of laziness. Despite the sinfulness of sloth, the King James Bible, as we have already mentioned, does not include the word 'sloth' at all, though there are two references to 'slothfulness' (Proverbs 19:15 and Ecclesiastes 10:18) and fifteen instances of 'slothful', all but four of which are in the Book of Proverbs. My favourite of those is Proverbs 26:14 which says: 'As the door turneth upon his hinges, so doth the slothful upon his bed,' which is not intended as a compliment, but still has something very restful about it. The following verse is also rather ambiguous: 'The slothful hideth his hand in his bosom; it grieveth him to bring it again to his mouth.' One could interpret that as seeing slothfulness as beneficial to dieting, but I don't think that was the intention.

More recently, sloths crept slowly into the awareness of novelists. Walter Scott, in his 1819 novel *The Bride of Lammermoor*, subscribed to a common uncomplimentary belief about sloths when he wrote: 'In short, Dick Tinto's friends feared that he had acted like the animal called the sloth, which, having eaten up the last green leaf upon the tree where it has established itself, ends by tumbling down from the top, and dying of inanition.' Charles Kingsley was far more pleasant in *Westward Ho!* (first published in 1855), which includes the line: 'Now and then, from far away, the musical wail of the sloth, or the deep toll of the bell-bird, came softly to the ear.' However, we had to wait until 2001 and Yann Martel's *Life of Pi*, for an unequivocally complimentary sloth reference: 'My zoology thesis was a functional analysis of the thyroid gland of the three-toed sloth. I chose the sloth because its demeanour – calm, quiet and introspective – did something to soothe my shattered self.'

Led by writings such as this, and the children's books mentioned earlier, it is fair to say that the twenty-first century brought in a cultural sloth renaissance. Even before this had taken root, however, there was one area which was always appreciated by adults with a sense of humour: sloths in newspaper and magazine cartoons, where their generally lethargic manner offered rich pickings. Here are my personal top ten captions to sloth cartoons, with the names of the cartoonists (in upside-down order, of course):

10. [A sloth in a tree talking to suited, managerial sloth busy at desk]: 'I just want you to know, you're a disgrace to every sloth in this jungle, and to slothdom in general.' (Veley)
9. [Two sloths hanging upside down]: 'I need a commitment. I don't want to just hang out with you.' (Bacall)
8. [Three sloths in tree looking disparagingly at two below them, dancing energetically on the ground]: 'It's some hormonal imbalance, apparently.' (Duncan)
7. [Two sloths in a tree]: 'I'll count to ten thousand. You go and hide.' (Hagen)
6. [Two sloths in a tree; two jumping around on the ground]: 'These teenagers... when will they learn to lie around and do nothing all day.' (Wilbur)
5. [Three sloths in a tree, one running on treadmill]: 'I'm on my break.' (Royston)

4. [Two sloths in a tree]: 'Had a go at that speed dating once... Wouldn't recommend it.' (Wilbur)

3. [Two sloths in a tree]: 'Sloth? We prefer Creature of Persistent Inactivity.' (Royston)

2. [Two sloths in a tree]: 'I don't know what's wrong with me today. I just feel so alert and full of energy.' (Royston)

1. [Two sloths in a tree]: 'Oh hell, are you sure? I was hoping we were Lust.' (Handelsman)

Animated cartoons have been a far less fertile breeding ground for sloths, at least until very recently. After being avoided by animators for ages, probably because they were relatively unknown animals, sloths experienced a further setback with the release of the 1985 non-animated adventure comedy *The Goonies*, which has as one of its main characters a deformed, very ugly and probably brain-damaged fellow called Lotney Fratelli, who is known to all as 'Sloth'. Even if he did end up as a hero at the end of the film, that left the reputation and image of sloths with much to overcome.

The *Ice Age* animated films tried to do so, starting in 2002, when the character of Sid the Sloth was introduced. Presumably, as this was the Ice Age, Sid is supposed to be a giant sloth, but his appearance introduces great doubt. He looks too big to be an arboreal sloth, but much too small and scrawny to be a megatherium. His very ugly face and ridiculously wide-spaced eyes also don't look much like a sloth.

I suppose we should put that all down to artistic licence, perhaps even arguing that in 2002, the sloth renaissance was still in its earliest days and many people didn't know what a sloth looked like anyway, but modern sloth-lovers will find it difficult to like Sid.

In 2013, the DreamWorks full-length animated film *The Croods* went a long way towards repairing the reputation of animated sloths. The story concerns a Pleiocene family, headed by Grugg and Ugga Crood (voiced by Nicolas Cage and Catherine Keener) and an inventive young fellow called Guy (Ryan Reynolds) who has a pet sloth called Belt (Chris Sanders). The name comes from Guy's habit of keeping the sloth around his waist to hold his trousers up, which Belt, as a very helpful and lovable sloth, is delighted to do.

Belt did a good job for sloths, but the animation that really lifted them was the Oscar-winning *Zootopia* from Disney, and as all sloth-lovers would agree, it was one scene in particular that won the film the Academy Award for the Best Animated feature. The scene takes place when the crime-fighting duo Judy the rabbit and Nick the fox urgently need to trace the owner of a car. They have written down the car's registration number, so they take it to Nick's friend Flash at the Department of Mammal Vehicles. But Flash, much to the consternation of the highly impatient Judy, is a sloth. A friendly sloth, a helpful sloth and a thoroughly well-meaning sloth, but a very, very slow sloth. And the deliciously slow animation of his slow-motion speech and movement is glorious.

Judy's anguish is further increased when Nick tells a joke to Flash who laughs at it very, very slowly, and then tells it to his colleague Priscilla. She, incidentally, in a perfect in-joke for sloth-lovers, is voiced by Kristen Bell, whose slothy meltdown we shall meet on p. 172. If there were an Oscar for Best Animated Sloth, however, it would without doubt go to Flash.

Zootopia was released in March 2016, and however large its impact was on sloth culture, that was nothing compared with the slothmania that was to follow. Up to that point, sloths had been unfairly accused of many things, from being badly designed to being lazy, from being melancholy to being a Deadly Sin. One sin nobody would accuse them of, however, is being self-publicists. Hiding at the tops of trees from predators on the ground, and hanging motionless for most of the time, while algae grow on their fur to disguise their presence from predators in the air, they did less than any other animal of comparable size to attract publicity. In the last few years, however, the sloth publicity machine has begun to take off.

First, came the glorious work of enlightened sloth researchers such as Bryson Voirin and Becky Cliffe, who not only discovered things about sloths that had never been known before but told the world what amazing creatures they really are; then came the videos, photographic collections and documentary films of Lucy Cooke, which revealed how beautiful and adorable sloths can be; then the big-time animators such as Disney and DreamWorks got in on the act with creations such as the Sid the Sloth in *Ice Age* and Flash, the delightfully slow sloth in *Zootopia*. This set the scene for the pandemic of slothmania that struck the civilized world in 2017, but even before that, there was another vital area which contributed to sloth awareness: the world of advertising. Being amazing and adorable is all very well, but what the advertising industry had begun to realize was the propensity of ordinary people to identify with the languid nature of sloths.

In 2013, Sofaworks (which later changed its name to Sofology) introduced an advertising campaign based on an animated sloth character named Neal, who liked nothing better than reclining on a sofa. The success of the campaign led to the introduction of Neal the Sloth soft toys, which came with an official 'Certificate of Slothenticity' to purchasers of sofas. By chance, the campaign coincided with the arrival at London Zoo of a baby sloth, the first ever to be bred at the zoo. The widespread press coverage of the baby (which zoo visitors voted should be named Edward, as its claws reminded them of Edward Scissorhands) linked in very well with the Sofaworks campaign, giving it a welcome boost, and other advertisers began to take notice, for Neal the Sloth was tapping into a sizeable section of the market which had previously been ignored.

The key to advertising had always been the manipulation of people's aspirations. Make them want something then sell it to them. Advertising was geared towards creating envy for high-energy, beautiful, trendy people, but what about all those who aspired to nothing greater than slumping contentedly on a sofa and watching the television? With global warming bringing energy efficiency into focus, a sloth was the perfect model to bring respectability to a low-energy, slothlike lifestyle. Remarkably, sloths were suddenly becoming fashionable.

This was not completely new. As long ago as 2000, Procter & Gamble had used a cartoon sloth to show the ease and joy of changing their Charmin toilet paper rolls, and in 2006 in Italy, a somewhat strange set of adverts for Alfa cars featured a rather unconvincing person in a sloth suit leading their 'Sloth speed: life on the slow lane' series; and the following year, an advert on US television for the Honda CR-V featured a sloth driving the car. However, we had to wait for Neal the Sofaworks sloth before a real fashion was set off.

In 2013, both Rockstar Energy Drink, based in Las Vegas, and Zlatopramen energy drink in the Czech Republic featured a sloth in their adverts.

A baby sloth will cling to its mother for most of its first year

Baby Lento at London Zoo beginning to show an interest in the world beyond mummy's tummy

Warm and safe: Lento peeks out from Marilyn's fur

Mixed veg for a baby two-toed sloth

Not lazy, just sleepy. Edward at ZSL London Zoo in 2015

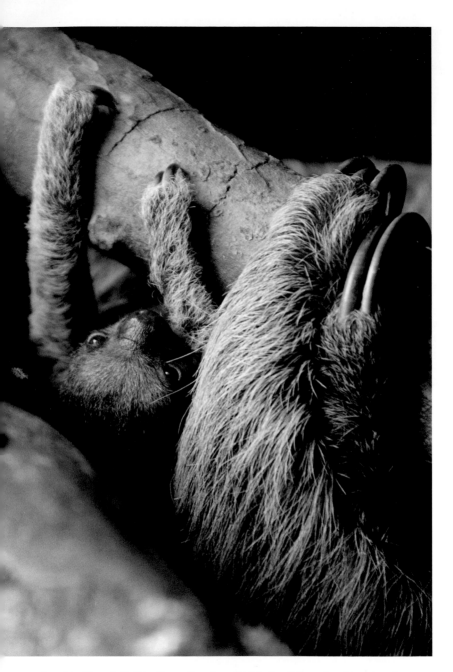

London Zoo sloth Marilyn gives her first baby, Edward, a hanging lesson

Do not disturb

We can be very fussy about the leaves we eat

Nisha Owen of ZSL calming a pygmy sloth before a backpack tracker is fitted

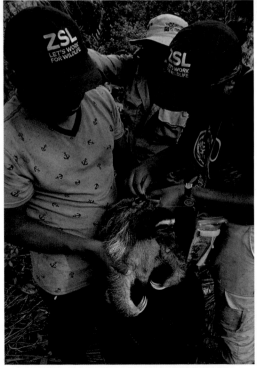

The sloth waits patiently as the tracker is attached

'Well, I'll do my best to keep it on, but I can't promise anything'

Pygmy sloth proudly displaying its tracking collar

'How sweet to be a sloth'

Sloth at sunset waving us farewell

In 2015, the urban clothing company American Apparel employed a sloth called Buttercup, the original inhabitant of the Sloth Orphanage in Costa Rica, to shift their image from risqué, nubile young women to a homely twenty-three-year-old sloth.

At around the same time, Mansur Gavriel used pictures of sloths to promote their upmarket handbag range, but the most unlikely use of a sloth in advertising came in 2017 with the creation of the 'dolph-a-sloth' for a streaming mobile phone service from Three UK. Combining the front half of a sloth with the back half of a dolphin, this creature exemplified both the laid-back features of the sloth and the energy and athleticism of the dolphin, perhaps to convey the idea of a couch potato leaping from channel to channel and film to film on the new device. The dolph-a-sloth became the face of the campaign for Three UK and a new dimension was added to sloth popularity.

Slowly but systematically, like a sloth climbing a very tall tree, sloths had worked their way to the top of the popularity jungle and in the run-up to Christmas 2017, they became a major theme, with sloths wearing Santa Claus hats among the simplest and most popular. There were wide ranges of sloth T-shirts, sloth jumpers, sloth wallpapers for computers, cuddly sloths, ugly sloths, packs of sloth Christmas cards, sloth gift-wrapping paper and most of the major supermarkets were even selling sloth calendars. Until then, the only Sloth of the Month calendars I knew were either the ones I had specially printed as personal Christmas gifts or those available to members of the Sloth Appreciation Society. Now every sloth fan could have them, along with every conceivable type of sloth merchandise including sloth mugs, sloth cufflinks, sloth duvet covers, sloth socks and even a perforated sloth tea infuser.

As a final piece of evidence of the worldwide boom in interest in sloths in 2017, I should mention that Guyana in that year issued a set of postage stamps depicting modern sloths to add to their giant-sloth stamp which had been issued in 1991. They were not the first country to show contemporary sloths on their stamps, however, having been preceded by Colombia (1960) and Nicaragua (1994), as well as two other countries where sloths had never set toe: Equatorial Guinea (1977) and Mozambique (2011).

Returning to the cultural history of sloths, we should mention the rather scanty appearances of sloths in music of which the British Library catalogue mentions only a handful. Most prolific among these were recordings by a garage rock band formed in Los Angeles in 1964 who called themselves The Sloths. They had reasonable success for a couple of years but disbanded in 1966 when one of their members left to study law. Their most successful single, 'Makin' Love', was rediscovered in 1984 and included in Volume 4 of the influential *Back from the Grave* compilations. In 2015, the group reformed and released their own *Back from the Grave* album to much critical acclaim.

Other musical sloths include Manfred Mann (on their 1972 *Earth Band* album), who recorded a track called 'Sloth', while Fairport Convention also featured a Sloth track on their 1970 *Full House* album, but the first of these was purely instrumental and the second did not include the word 'sloth' at all in the lyrics. Brendan Perry's song 'Sloth' (from his *Eye of the Hunter* album in 1999) also did not contain the word 'sloth'.

This leaves us with 'Son of Sloth' and 'Bride of Sloth', both by Stephen Dray, 'The Sloth and the Greed' by Pete Morton,

'Slobodan the Sloth' (from a group called Summon the Octopi) and 'Slow Hand Sloth' by Anthony Phillips, none of which I have ever heard.

On the classical side, the composer Elisabeth Lutyens was commissioned to write a choral work entitled 'Sloth' for a Seven Deadly Sins series, performed by The King's Singers at the Cheltenham Festival in 1974. This is a sextet for unaccompanied voices comprising a countertenor, two tenors, two baritones and a bass. A review in the *Spectator* rated this the best of the musical sins, describing it as a 'slithery evocation'. I strongly suspect, however, that it was not in the key of D sharp major, so would have been unlikely to attract the attention of sloths if William Beebe's comments (see p. 94) are correct.

The best Sloth song of all, however, remains the one written by Michael Flanders and Donald Swann for their *Bestiary of Flanders and Swann* album in 1967. 'A Bradypus, or Sloth, am I,' it begins, 'I live a life of ease', then continuous at a suitably languorous tempo, to tell of its leisurely lifestyle and the joys of being upside down while contemplating, at a much-increased speed, all the things it might do if it were not a sloth. The joys of swaying in the breeze 'suspended by my toes', however, take up all its time and transcend any alternative achievements, all summed up in the closing words 'How sweet to be a sloth'.

Let us end this chapter on sloths' appearances in culture with one of the most ancient sloth stories of all, a fable by Aesop which may date back to the sixth century BC. This version is from the edition published in 1818, collected by, and featuring the engravings of, Thomas Bewick:

Industry and Sloth

An indolent young man, being asked why he lay in bed so
long, jocosely and carelessly answered, 'Every morning
of my life I am hearing causes. I have two fine girls, their
names are Industry and Sloth, close at my bed-side as soon
as ever I awake, pressing their different suits. One intreats
me to get up, the other persuades me to lie still; and then
they alternately give me various reasons why I should rise,
and why I should not. This detains me so long, (it being
the duty of an impartial judge to hear all that can be said on
either side) that before the pleadings are over, it is time to
go to dinner.'

Of course, the subject of this fable is sloth the sin rather than sloth
the animal, but the story is so similar to the Karaja tribe's myth
about sloths given on p. 114 that I think it is worth a languorous
mention.

Finally, before we leave Aesop, another Sloth deserves a mention.
In 1840, the Canton Press published a translation into Chinese of
Aesop's Fables. It was not the first attempt to translate the stories
into a Chinese dialect, but was the first to gain general respect from
the Chinese as a faithful and idiomatic version. The translation
was specifically designed to encourage English readers to learn
and study Chinese and the title page explains that the fables were
'written in Chinese by the learned Mun Mooy Seën-Shang and
compiled in their present form (with a free and literal translation)
by his pupil Sloth'.

The man hiding behind that pseudonym was Robert Thom, an
English merchant who had studied the Chinese language while

working for Jardine, Matheson and Co. in Canton in the late 1830s. He became an official translator for the British during the first Opium War and was later appointed consul at Ningpo. I have no idea why he called himself 'Sloth', but he was described by one of his commercial employers as 'very industrious and obliging' which suggests that he must have been one of the least slothful Sloths of all time.

Finally, my favourite sloth joke:

A sloth was walking down a road when it was set upon by a gang of snails which beat him up terribly and left him injured and bleeding in the street. When the sloth had recovered enough, he crawled to a police station to report the assault.

'Did you get a good a look at your assailants?' the police officer asked. 'Can you give a description?'

To which the sloth replied, 'It all happened so fast.'

A misunderstood nineteenth-century British view.
How the English botanist and zoologist George Shaw
saw sloths in his 1808 work on mammals.

Chapter 14

ODDMENTS

**Thick hung the rafters with lines
of pendant sloths.**

Herman Melville, *Mardi, and a Voyage Thither*
(1849)

In the previous chapters, I have done my best to organize everything I have been able to find out about sloths into behavioural, physical, historical and cultural categories. This chapter is a potpourri of everything that is left over, in no particular order.

SLOTHS AND POLITICS

One of the rare references to sloths in the UK press before 2000 related to a decision taken in October 1999 on behalf of the Cabinet Office. Mo Mowlam, who had previously been Northern Ireland Secretary, had been moved to the job of Minister for the Cabinet Office and promptly made a deal with London Zoo to sponsor a sloth at a cost of £35. Announcing that 'The two-toed sloth was one of the most interesting species to choose from at London Zoo – but my choice is no reflection on members of the Cabinet Office,' she neverthless came in for some ribbing from *The Times* newspaper.

The sloth, *The Times* pointed out, 'spends most of its life asleep, hanging upside-down in trees, and takes a month to digest its high-roughage diet of leaves, twigs and fruit', going on to suggest that 'Dr Mowlam may have hoped to promote such "healthy" eating as head of the Department of Health, a job she would have preferred.'

Seeing that report encouraged me to find out whether sloths had ever been mentioned in House of Commons debates, so I looked through the archives of Hansard, the official parliamentary record. The first reference I found was from a debate held on 31 July 1833 on the state of convict ships when Daniel Harvey, the MP for Colchester, made this uncomplimentary and incorrect remark about sloths: 'The law might be said to cling to this estate, as the hungry sloth clings to a luxuriant tree, and after battening upon it until scarcely a leaf was left, had fallen upon the grass in a bloated stale of repletion.'

Apart from that, all my search of Hansard revealed seemed to be an amusing contest between two recent Labour Party MPs to see who could mention sloths more often.

Paul Flynn, the Member for Newport West, kicked off on 4 June 1991, in a debate on the Child Support Bill, when he said this: 'As always, the Government behave with energy and speed – the speed of a striking cobra – in pursuit of those whom they believe are gaining very small sums, but when it comes to pursuing the large-scale scroungers in our society, the income tax evaders and avoiders, they move with the speed of an arthritic sloth.'

Only a few weeks later, on 27 June 1991, Frank Dobson, Member for Holborn and St Pancras, equalized by referring to the difficulty in getting known drug dealers out of blocks of flats, saying, 'The system is so slow that it makes a two-toed sloth seem hyperactive.' Quite why he specified two-toed sloths is unexplained, for three-toed sloths are even slower, but perhaps the comment was based on a precise comparison between the Choloepus, the Bradypus and the lethargy of drug-dealer evictions.

The score stayed level at one-all for almost five years, until 10 January 1996, when Dobson took the lead by saying, when talking about the establishment of enterprise zones: 'The Government's record in that area has shown that they have acted slower than a two-toed sloth.'

Flynn equalized, on 10 July 1997, in a debate on the mis-selling of personal pensions, with a replay of the joke he had made six years earlier: 'They move with the speed of a striking snake on speed when selling pensions, but when it comes to compensation their reactions are those of an arthritic sloth.' This must have gone down well, because Flynn used the same analogy twice more: on 9 February 2012, when he asked, 'Why do the Government move with the speed of a striking cobra when they are slashing support for essential renewable energy, but with the speed of an arthritic sloth when it comes to recognizing the subsidies that will be essential for nuclear power in future?' and on 4 June 2013, when he again asked, 'Why do the Government move at the speed of a striking cobra in further impoverishing the already poor with the bedroom tax, and why, in the case of reforming the parasitic incubus on the body politic of lobbying, do they move at the speed of an arthritic sloth?'

With Flynn now leading 4-2, Dobson may have realized that he was beaten, for he stood down as an MP at the 2015 General

Election. Interestingly, the legitimacy of Flynn's repeated concern for arthritic sloths was confirmed in an article on the American HealthChiro web site in 2014 entitled 'What Do Bats and Sloths Have To Do With Osteoarthritis?' The writer firmly asserts that: 'The myth that these creatures don't show wear and tear in the joints despite being vertebrates like the rest of us, comes from the idea that the massive amount of hanging about that they do, eliminates the compressive forces of gravity. While it may decrease it, it probably doesn't eliminate it, so sorry folks, poor bats and sloths age the same way we do. Maybe the degree of the wear and tear is less extensive for them, but they're not exempt from gravity.' So arthritis can be a problem for sloths, and sloth-kind should be grateful to Paul Flynn for drawing attention to it.

The only other connection I have come across between sloths and politics came in 2013 when a glorious collection appeared on the Internet of '21 Sloths That Look Like Ed Miliband'. This featured twenty-one pictures of sloths placed alongside twenty-one pictures of the then Labour leader, in each of which he was wearing an expression remarkably similar to that of the neighbouring sloth. In some cases, in fact, the match was so uncanny that Miliband almost looked more like a sloth than the sloth.

Despite this, Miliband's Labour Party lost the 2015 General Election. Perhaps the British people were simply not yet ready to have a sloth lookalike as their Prime Minister.

SLOTHS AND HERALDRY

The Society for Creative Anachronism has always taken very seriously its duty of approving designs for heraldic badges and coats of arms in America, even if they lack the due authority bestowed by the College of Arms in London, and the Lyon Court in Scotland, which do a comparable job in the United Kingdom. The SCA have in any case been caused some problems by the sloth.

In July 1996, they granted an application for a 'Device. Argent, a sloth pendant contourney vert from a rod fesswise sable' to an applicant called Sven Örfhendur. His shield displayed a rather muscular green sloth hanging upside down from a black rod. By March 2017, however, they had clearly had second thoughts, as expressed in a reply to Sorcha MacKenzie, who had applied for a 'Badge (Fieldless)' consisting of 'A sloth argent charged with a tau cross sable, pendant from a rapier fesswise Or'. An earlier application had been turned down and returned for redesign, but even though the new application addressed the problems raised, it was still found to fall behind the high standards demanded by the SCA: 'Despite addressing the reasons for the previous return, this badge must be returned for redesign,' they wrote:

> The use of a sloth, as a New World animal, is a step from period practice. Its posture, comparable to the only regis- tered instance of a sloth (Sven Örfhendur, 7/1996), was then blazoned 'pendant'. However, while descriptively accurate, this is not an heraldic posture for a quadruped. At best, its posture can be described as 'statant inverted, pendant from...' However, under current precedent, beasts inverted are unregisterable except for limited circumstances, such as bats, which get a step from period practice... At best, this would receive a second step from period practice, making it unregisterable. At worst, its inverted posture makes it unreg- isterable even without the existing step from period practice.

Occasionally, upside-down bats (or 'reremice' as the SCA anach-ronistically called them) had previously been allowed, but a ruling in February 2017 announced that this would be forbidden from September 2017. In its general rules, the SCA states that: 'The College's standards have changed dramatically over time. Newer rulings override old ones, so rulings appearing in old letters do not necessarily indicate the current policies and practices of the College.' Upside-down sloths may have been OK in 1996, but in 2017 they failed on two counts: upside-downness and being American. And even European bats now have to be the right way up.

SLOTHS AND DETECTIVE STORIES

There is, for some lamentable reason, a distinct shortage of sloths in detective stories. In fact, I have found only one such story in which a sloth plays a major part. Published in the *Baltimore Sun* newspaper in 1912, it was called 'The Robbery at the Tower', was written by Howard Carter and was subtitled: 'An Adventure of Peter Crewe – The Man with the Camera Eyes'. The story concerns the theft of some famous jewels from the Tower of London:

> They were kept in an isolated building, a round turret which ran straight up to a height of seventy feet, and was absolutely inaccessible from the outside, the brick walls affording not the slightest foothold... This turret had a small barred window overlooking the road, 60 feet up. It was too high for any thief to throw up a grappling hook... Yet in spite of this, the window was entered from the outside, a bar was removed from the mortar setting, and the thief gained entrance, obtained posses-sion of the emeralds, and calmly descended, unobserved.

The only suspicious character seen in the vicinity was an Italian organ-grinder with a sick monkey, but that was enough, together with an examination of the wall, for the Sherlock Holmes-like Peter Crewe to solve the crime. The vital clue was scratch marks like hen-tracks on the wall leading up to the window.

SPOILER ALERT! This was enough for the detective to come to the inescapable conclusion that the organ-grinder was not Italian but South American and his 'sick monkey' was not a monkey but a sloth. As Crewe explains, 'I saw its marks upon the brick work of the tower... The sloth has only three toes, and its marks are as much like hen-tracks as anything in the world.' Furthermore,

> Is it not reasonable, therefore, to assume that the man who owned it was intimately acquainted with it – in other words that he was a denizen of South America, the continent in which the sloth has its habitat? And if that deduction were not logical enough, we have the testimony... that the man was an Italian – the generic term in England for all southerners of swarthy complexion.

He goes on to explain that a long coil of rope was attached to the sloth's body and the animal placed upon the wall. 'Our sloth, feeling the smooth surface beneath him, and being unable to sleep perpendicularly, conceives the idea that he is upon the stem of a particularly high and smooth palm tree, at whose summit he may hope to find a comfortable branch from which it is his delight to feed.'

So the sloth climbs the wall and when it comes to the window bars he imagines they are branches and grips them tight enough for the thief to be able to climb the rope and falls asleep. Quite how the sloth climbs a perfectly smooth wall while making only a few scratch marks is not explained, but Crewe's reconstruction is confirmed when he spots a swarthy fellow carrying a suspicious-looking bag in the Spanish quarter of Bloomsbury ('These people would rather die than live outside their own neighbourhoods') and follows him to the scene of his next robbery. He watches the sloth being taken from the bag to climb the wall of a hotel, whose rooms promised rich pickings. When it reached a window, the man started climbing the rope but panicked when Crewe shouted 'Halt, or I fire.' The rope then began to slip from the man's waist and ended up strangling him. The sloth, I am happy to say, lived out its days peacefully in the Zoological Gardens.

Just think: if the fictional Peter Crewe had only been called upon to examine the mysterious Brazilian tunnels we mentioned on p. 32, he might have deduced at once that they were dug by giant sloths.

SLOTH INTELLIGENCE

Researchers have yet to come to grips with the matter of the mental abilities of sloths and the question of the extent to which they can be trained to do things outside their normal repertoire. Goffart's *Function and Form in the Sloth* includes one reference to unpublished research by D.B. Lindsey and R.B. Livingstone into the ability of sloths to learn a maze. Apparently they learned the correct route in two or three days. Then came the real test of mental agility. 'After they had learned the maze without error, a blockade was introduced with the idea of frustrating them. A new pathway to the goal was also introduced,' but it was the researchers who ended up most frustrated. They report: 'Two specimens struggled trying to get around the obstacle and after trying all conceivable means showed manifestations of internal inhibition in the form of sleep.'

In other words, the sloths, after making some sort of effort, thought: 'Sod this for a game of psychological experimentation. I feel like a spot of energy conservation.' The sloths' ability to learn the maze in the first place, however, was thought to be equal to that of an average dog or cat.

The 'manifestations of internal inhibition', however, are rather reminiscent of a passage in William Beebe's 1925 account of an unsuccessful mating attempt by one male sloth. It began promisingly with the male getting his claws entangled in the fur of a sleeping female, but problems arose when the female awoke:

> When she grasped the situation, she left at once and clambered to the highest branch tip followed by the male. Then she turned and climbed down and across her annoyer, leaving him stranded on the lofty branch looking eagerly about and reaching out hopefully toward a big green iguana asleep on the next limb in mistake for his fair companion. For an hour he wandered languidly after her, then gave it up and went to sleep.

Both these examples suggest that sloths are adopting a very sensible motto: *If at first you don't succeed, give up and have a nap*. The picture of sloths losing interest in problem-solving is completed by a news report in 2007 concerning a sloth named Mats which stubbornly refused to take part in experiments at a university in Germany. Mats was hailed in some accounts as the laziest animal in the world after the following story appeared on the AP news wire:

> Scientists in the eastern German city of Jena said they have finally given up after three years of failed attempts to entice a sloth into budging as part of an experiment in animal movement. The sloth, named Mats, was consigned to a zoo after consistently refusing to climb up and then back down a pole as part of an experiment conducted by scientists at the University of Jena's Institute of Systematic Zoology and Evolutionary Biology. Mats was not even tempted by cucumbers or plates of homemade spaghetti.
>
> 'Mats obviously wanted absolutely nothing to do with furthering science,' said Axel Burchardt, a university spokesman.

On a more positive note, keepers at London Zoo report success in training their young sloth Edward to enter a box, prior to its travelling to Vienna to meet a girl sloth in 2015.

Back in Germany, however, we should mention that when the uncooperative Mats was sent in disgrace to the zoo, his replacements, three two-toed sloths called Julius, Evita and Lisa, did the researchers proud in their research into sloth locomotion. The idea, which Mats had found so unappealing, was to use X-ray video equipment to analyse the motion of sloths walking along a pole in an X-ray tube. 'To our great surprise the locomotion of the sloths is basically not so different from the locomotion of other mammals, like monkeys for instance, which instead of hanging from tree branches, balance along them,' said Dr John Nyakatura, who led the research team. In other words, the walking motion of sloths is much the same as that of other animals, only upside down. What was most remarkable was that the sloth's long arms, very short shoulder blades and narrow, rounded chests all combined to give them the flexibility that leads to maximum energy conservation as they walk, and this ideal design had evolved separately in two-toed and three-toed varieties.

SLOTH CREATIVITY

With such a simple and predictable lifestyle, sloths are not generally called upon to come up with new ideas, but there is one piece of human creativity for which the inspiration has been ascribed to sloths. This was explained in a short article in *Pearson's Magazine* in 1900:

> There is one animal which lives entirely in trees, but is able to maintain its position during slumber without the least exercise of muscular force. This is the sloth, common in the forests of tropical America. Its long claws are so bent that they hook over the branches and allow the creature to hang upside down like an animated hammock. Curiously enough, the hammock appears to be a South American invention, and is usually employed by all the Indian tribes of the Amazons. Perhaps the primitive human dwellers in this region took to sleeping in hammocks after observing the habits of the sloth.

The article gave no further details, but support for its claim may be provided by the *Oxford English Dictionary*. The word 'hammock', according to the dictionary's earliest citations, was first recorded in English in 1555 in a work entitled *The Decades of the Newe Worlde or west India conteynyng the nauigations and conquestes of the Spanyardes* which was originally written by the Italian historian Peter Martyr d'Anghiera and translated into English by Richard Eden. He wrote of 'Theyre hangynge beddes whiche they caule *Hamacas*'.

In 1596, writing of 'The discoverie of the large, rich, and bewtiful empire of Guiana', Walter Raleigh said, of the inhabitants: 'They lay each of them in a cotten Hamaca, which we call brasill beds.' With Brazil and Guiana both sloth-rich countries, perhaps it is not so far-fetched to suggest that the natives got the idea of hammocks from the creative sleeping habits of the sloth.

SLOTHS AND GLOBAL WARMING

In April 2017, an unusual paper appeared in the journal *Mammal Review* entitled 'Green Sloths and Brown Cows: The Role of Dominant Mammalian Herbivores in Carbon Emissions for Tropical Agroecosystems', written by Jonathan Pauli, Zachariah Peery and Cayelan Carey. Thoughtfully argued, it contained a thoroughly pro-sloth account of the potential contribution of sloths towards saving the planet from the excesses of global warming.

Their argument was essentially an assessment of the carbon emissions of sloths compared with those of cows, and the reason for making that comparison was the rapid increase in dairy farming and beef production that has been taking place in Costa Rica. 'In many tropical forests of Central America,' they point out, 'tree sloths (suborder Folivora) are the dominant mammalian herbivores.' This dominance, however, is being lessened by the amount of deforestation taking place to create room for cattle to graze. Simply put, this has the effect of replacing sloths with cows.

The energy efficiency of sloths, however, leads to their being responsible for remarkably low carbon emissions, and these are made even lower by the algae living in their fur, which are net consumers of carbon. According to the calculations of Pauli, Peery and Carey, the average sloth emits only 12 grams of carbon a day, while the average cow emits a massive 2.3 kilograms.

Furthermore, sloths are solitary animals and tend to live well apart from other sloths while cows live in herds. So despite the difference in size, there tend to be far more cows than sloths in an acre of land where they are kept. The researchers estimate that replacing sloths with cows is leading to the emission of an extra 166 tonnes of carbon annually. They do not go so far as to suggest that all the world's cattle should be replaced by sloths, though some might argue that quite apart from the effect on global warming, it might leave the world a more peaceful place.

SLOTH PHILOSOPHY

In 1974, the American philosopher Thomas Nagel (b. 1937) raised the unanswerable question, 'What is it like to be a bat?' As far as I know, no philosopher has raised the same question about sloths, but two writers have given opinions on it. William Beebe began his generally sloth-friendly writings about the animal with these words:

> Sloths have no right to be living on the earth today; they would be fitting inhabitants of Mars, where a year is over 600 days long. In fact, they would exist more appropriately on a still more distant planet where time – as we know it – creeps and crawls instead of flies from dawn to dusk.

The palaeontologist, zoologist and science historian Stephen Jay Gould (1941–2002) developed the idea of the sloth as an unknowable, unearthly creature in his book *Leonardo's Mountain of Clams and the Diet of Worms* (1998):

> Sloths move with such pervasive slowness that their entire world seems intrinsically and permanently different from ours. I would almost conjecture than a fixed slow-motion camera occupies their cranial space, and that they gauge all their movements by this markedly different clock upon the world. Do we, and most other creatures, appear to them like the Keystone Cops in movement, or the Munchkins in raised pitch? Or do our frenetic paces (compared with their stately step) constitute the only external world they know, recorded in their brains as the slothful equivalent of 'objective reality'?... But sloths do know other sloths and must also perceive a differently paced external world as well. Perhaps they don't notice the difference; perhaps they are merely amused; perhaps they don't care. I would love to know. In any case, philosophical speculation aside, I have never been so powerfully moved by a sense of pervasive difference for something so basic as a pace of life.

Chapter 15

COSTA RICA

We will now close our account of the misnamed
sloth, by hoping that our readers agree with us in
thinking, that so far from its being the indolent
monster of deformity it has generally been represented, that it is one amongst numberless examples
which might be particularly selected as an instance
of Divine wisdom and beneficence.

Wonders of the Animal Kingdom (Society for
Promoting Christian Knowledge, 1847)

Costa Rica is the only suitable topic for this final chapter, because
that was where my love of sloths began, not from a visit to the
country, but from a wealth of glorious YouTube videos filmed at
Judy Arroyo's Sloth Sanctuary in Limon Province. Often referred to
as a 'sloth orphanage', its role in popularizing the image generally
held of sloths can hardly be overestimated.

Their first orphan, Buttercup, incidentally, is the oldest sloth
whose age has been definitely confirmed. Research on sloths in the
wild is so recent that their natural lifespan is not yet known. The
figure of 20–30 years is often quoted, but that is really little more
than a guess, combined with the example of Buttercup.

The fame and influence of the Sloth Sanctuary, however, was mainly local until 2012 when two English zoologists and one American actress were between them responsible for turning sloths into an international phenomenon.

First came Lucy Cooke, zoologist and film-maker, who saw a video from the Sanctuary which she described as 'the funniest thing I had ever seen', so promptly set off to Costa Rica to film a documentary called *Meet the Sloths*, which appeared in 2013. Even before that, to spread the good word about sloths, Cooke had founded the Sloth Appreciation Society at the end of 2011 and its membership worldwide is now around 8,000. Before the documentary came out, as a taster, she posted a short video on Vimeo and YouTube which rapidly went viral: 400 views on the first day, 40,000 the following day, 150,000 the day after that. Those figures, however, were outdone by the American actress Kristen Bell, who appeared with American comedian and chat-show host Ellen DeGeneres on the *Ellen* show in January 2012 to tell of the thirty-first birthday present she received from her boyfriend.

Kristen was another person who had begun a love affair with sloths after seeing a YouTube video from Costa Rica, but she suspected nothing when she had been told that she would be receiving a birthday present that no one else in the world would get. Told that it was in the next room, she entered and found herself face-to-face with a sloth. Her shrieks of joy and emotional meltdown were filmed and shown on the chat show as well as being posted on the Internet. In one day, it received two million hits, and the last time I looked was over 26 million.

To complete this trinity of sloth high priestesses, I must mention Becky Cliffe, who studied zoology at Manchester University then went on to Swansea to specialize in sloths for her PhD, centred on research in the rainforests of Costa Rica. Quite apart from writing several influential papers, which we have already referred to, on the anatomical and behavioural wonders of sloths, she played a large part in Lucy Cooke's *Meet the Sloths* documentary, as well as founding the Sloth Conservation Foundation in 2016 and doing more than anyone to spread knowledge and understanding of these previously misunderstood creatures.

After being converted to sloth-worship myself by this Lucy/Kristen/Becky trio, and finding out everything I could about sloths and writing the previous chapters of this book, I decided it was time I met some sloths in the wild myself. Apart from meeting, at a distance, London Zoo's resident two-toed sloth Marilyn, I had never seen a living example, so I booked a trip to Costa Rica to see what I could discover. Rather than heading to the famous Sloth Sanctuary, which is situated near the Caribbean coast of Costa Rica, I went to the other side of the country, to a hotel on a beach by the Pacific.

I quickly discovered that Costa Rica had far more to offer than just sloths. Quite apart from the luxury of a hotel on the beach and morning swims in the sea before breakfast, I enjoyed the company of an iguana or two by the hotel swimming pool then saw some splendid vultures and a toucan in the trees. But no sloths.

On my second day, I watched the howler monkeys in the trees on one side of the hotel and the capuchin monkeys on the other side. By that time, I was wishing I had an 'I-Spy Costa Rican Wildlife' book in which I could tick off the various animals. But the sloths, both Bradypus and Choloepus, would still have been unticked. I had, however, booked my place on two sloth-finding expeditions, so was still confident.

On the first of those expeditions, we almost spotted a crocodile, which was pointed out by the coach driver, but he was unable to stop the vehicle on a bridge. Then we reached a wildlife conservation area which we were assured had a healthy sloth population, and that's where I discovered just how difficult it can be to see the sloths in the trees. They may come down once a week to go to the toilet, but they certainly didn't do so while I was there, and while they are near the top of trees in the rainforest, they can be extraordinarily difficult to spot.

The sloth-spotting guides are amazingly good at finding them, but when they were pointed out to me, I still had great difficulty locating them, even when looking through the guide's carefully and accurately focused binoculars. I was, however, somewhat reassured about the value of my research when I was able to correct, on two or three occasions, the sloth-related information the guides were giving.

My second sloth-finding tour was more successful, with several views of far-off sloths climbing through the upper branches of Cecropia trees, and offered a bonus of some gorgeously coloured butterflies and frogs, but I was still denied the joy of a proper sloth close-up.

I was finally able to savour this when making a trip to the Diamante Eco Adventure Park and Animal Sanctuary, just round the corner from my hotel, where I arranged my arrival perfectly for the sloth feeding time. All three sloths were two-toed Hoffmann's Choloepus, which made them easier to feed than the fussy, Cecropia-munching Bradypus, but as the sloth-feeder Adriana told me, that did not stop them having clear taste preferences too. Gilbert, the little boy sloth, would only eat green beans, while Mia, larger and female, ignored the beans, preferring carrot and sweet potato. The third sloth in the enclosure, however, was Adriana's favourite, and as far as I could see liked nothing better than lying on her back in the sunshine. Her name was Lucy, though I think that there was no intended connection with Lucy Cooke, founder of the Sloth Appreciation Society and its supposed Headquarters in Slothville.

On my last sloth-finding tour in Costa Rica, I decided to find out what the locals really felt about sloths. Did they revere these animals, either for their individualistic behaviour, or their undoubted credentials as a major tourist draw to Costa Rica? Did they realize that sloths are, as one recent researcher put it, masters of an alternative lifestyle? Did they see them as one of nature's finest examples of perfect evolution for hanging upside down in trees and conserving energy?

'What do Costa Ricans generally think of sloths?' I asked. 'Do you praise them as Costa Rica's unofficial national mammal? How do you see them?'

He thought a moment, then gave his considered response: 'Lazy bastards.'

As I said right at the start, I blame the Portuguese for calling them 'lazies' in the first place.

Further Reading

Rather than compile the usual list of the many books and academic papers I have consulted in writing this work, I think it will be more useful to offer a guide to the best places to look for more information about sloths.

First, the books dedicated to sloths alone:

- *Function and Form in the Sloth* (1971) by M. Goffart was for forty years the essential work on sloths, giving a detailed summary of all that researchers had found out about them. Dry and academic in style, but full of wondrous information.
- *A Sloth in the Family* (1967) by Hermann Tirler is a short but delightful account of the wonders of having a pet sloth. Pre-empting slothmania by half a century, Tirler was perhaps the first to truly appreciate the Joy of Sloth.
- *Sloths: Life in the Slow Lane* (2017) by Rebecca Cliffe and Suzi Eszterhas is a glorious collaboration between one of the greatest sloth researchers and one of the finest wildlife photographers to give a beautifully illustrated insight into the secret life of sloths.
- *Life in the Sloth Lane: Slow Down and Smell the Hibiscus* by Lucy Cooke (2018) is a magnificent collection of photographs of sloths, interspersed with inspiring quotations promoting the glories of a slothful existence.

Next, a work that devotes only one chapter to sloths but comes closer than anything that came before in capturing their special qualities:

- *The Unexpected Truth About Animals* (2017) by Lucy Cooke is a hilarious romp through a zoologist and film-maker's experiences with over a dozen animals, with wondrous revelations about each of them; especially, as one would expect from the founder of the Sloth Appreciation Society, sloths.

Moving away from books, the Internet has played a great role in popularizing sloths, starting with the videos on YouTube from the Sloth Orphanage in Costa Rica which taught the world how adorable these animals can be. If you are in search of reliable, up-to-date information about sloths rather than just lovable images, I can recommend the following sites:

- www.slothville.com, which is the home page of the Sloth Appreciation Society and offers much sloth information and merchandising.
- beckycliffe.com, where the world's foremost sloth researcher keeps us in touch with her work and her continuing passion for sloths.

The must-see films for the sloth addict are these:

- *Zootopia* (2016), *Ice Age* (2002) and its sequels, particularly *Surviving Sid* (2008), and *The Croods* (2013) are all animations with high sloth content, but *Zootopia* is the only one that captures their essentially adorable nature.
- *Life of Pi* (2012) features a real, live sloth right at the start, but it makes no further appearance. Read the book for a proper respectful treatment of sloths.

SLOTH BISCUITS

Of all the sloth-related paraphernalia I have acquired or created through my fascination with these animals, the ones that have earned the greatest acclaim have been my sloth-shaped biscuits. So here's how to make them.

First, of course, you need a sloth-shaped biscuit cutter, which is something that is not as easy to come by as it ought to be. In fact, an Internet search led me only to cookie cutters from the USA for which the postage was four or five times as much as the cost of the cutter.* For someone as parsimonious as myself, that seemed outrageous, so I resolved to make my own, which bring us to Stage One in the sloth biscuit programme:

Making Your Own Sloth Biscuit Cutter

1.   Purchase a cheap tin of tuna (or other food in a slim tin).
2.   Cut top and bottom off tin and throw away the tuna to leave a cylinder of metal. Alternatively one may use the tuna to

---

* I have since been sent a sloth cookie cutter by a friend in the United States. I am most grateful, but must say it is not as elegant as my home-made versions.

make fishcakes, but it is scarcely worth the trouble. Tinned tuna is really not very nice.

3. With the help of a pair of pliers if necessary, depending on the firmness of the metal, bend the cylinder into a sloth shape.

TIPS

a) It may help to sketch out a template for the sloth shape on a piece of cardboard then measure round its perimeter to ensure it matches the circumference of the tin. If it doesn't match, amend the template.

b) Tuna tins seem to have the right metal consistency to make biscuit cutters. Some tins are too substantial to bend easily, even with the help of pliers, others are too flimsy and split when bent. This can be dangerous.

You now have your biscuit cutter, which brings us to Stage Two.

Making the Biscuit Dough

My long experience of biscuit-making has led me to an all-purpose recipe. You will need:

150g plain flour
120g unsalted butter
90g golden caster sugar
1 tbs maple syrup

You can then add other ingredients to make a wide variety of flavours. Mix by hand and this will be enough for around 24 biscuits.

SUGGESTIONS

Dark chocolate chips to add flavour.

Dried orange (or even better clementine) peel reduced to dust in spice grinder with the seeds of half a dozen cardamom pods to add some class to the chocolate chip version.

One level tbs of ground ginger plus half a tbs of cinnamon for ginger biscuits.

Seeds and juice of one passion fruit (omit seeds if fussy) plus chopped white chocolate for delicious fruity biscuits.

Roll out pastry and use your home-made cutter to cut out the slothy shapes and you are now ready for the optional final touch in the biscuit formation.

ATTACHING SLOTHS TO TREE BRANCHES

This final touch happened by chance in a piece of pure sloth serendipity when I was making some sloth ginger biscuits. After cutting out the biscuits, I had, as usual, some leftover dough. Instead of my normal practice of using this up by making a circular biscuit or two, I rolled it out and sliced it into strips to create ginger thins. (These are delicious when stuck in my home-made lemon and honey ice cream, incidentally.)

I cooked the sloths and the thins together in the oven and was amazed when I removed them at the end of the cooking time to find that one of the sloths had attached itself to a thin strip and looked just like a sloth hanging upside down from a tree. Since then, that has become my standard design, as the picture on p. 183 shows.

Attaching the sloth to the branch demands some care, especially if the dough is dry, but it is well worth the effort. I find that the point of a knife comes in very useful for softening up the sloth's feet, then blending them into the branch. You are now ready to put the cookies in the oven:

Cooking the Biscuits

- Bake at 180°C for 13 minutes (11–12 minutes if you want them chewy inside, 14–15 minutes for crunchy biscuits).
- Remove from oven and leave to cool.

You are now ready for the final touch.

FINALLY…

1. For chocolate biscuits, melt some chocolate and smear over one side of the sloths (optional).
2. Dunk entire biscuit in melted chocolate for even more chocolatey biscuit (equally optional).
3. For absolute slothiness: melt chocolate and, using a kebab stick, or cocktail stick, or other pointed object, draw sloth faces and claws on the biscuits. If you've coated the biscuit in chocolate, make sure it is dry before adding these items. Use white chocolate for the faces if the biscuits have already been dunked in chocolate; use dark chocolate otherwise. I always draw three-toed biscuits, but two-toed is also acceptable.

A sextet of hanging sloth biscuits around an inlaid sloth.

FOR A TRULY ARTISTIC VERSION:

- Roll out dough thin, and cut into equal-sized rectangles slightly larger than your sloth biscuit cutter.
- Cut out sloths from the centre of half the rectangles.
- Place the remainder of each rectangle of dough on top of a complete rectangle and press down slightly. This produces biscuit shapes with sloth-shaped holes in them.
- Bake these biscuits for the usual 13 minutes.

- Remove from oven, leave to cool, then pour melted chocolate to fill the sloth shapes. When chocolate has set, add faces and claws as usual.

Result: beautiful, edible framed sloths.

Acknowledgements

So many people have helped and encouraged me in writing this book that I do not know where best to start this catalogue of thanks. First, perhaps, I should mention the two people whom I see as the High Priestesses of the cult of sloth worship. Right at the start of my research, I had glorious, hour-long telephone conversations with both Rebecca Cliffe and Lucy Cooke, who between them have turned sloths from derided, or at best ignored, creatures into one of the world's most adored animals. Lucy's glorious films and Rebecca's research have completely changed people's attitudes towards sloths, without which I would never have become fascinated with sloths in the first place. Their infectious humour and enthusiasm for the subject, and their welcoming encouragement in venturing onto their territory was a great launch for the project.

Having started on my sloth expedition, and watched all Lucy's glorious videos on YouTube, my next stop was London Zoo to meet a real, live sloth and I am hugely grateful to everyone at that wonderful institution for facilitating my visits and making them such a pleasure. Sandra Crewe introduced me to their resident sloth Marilyn, with whom I fell in love at first sight. Marilyn even moved while I was watching her, which I was told was a great compliment. My research in the Zoo's archives was greatly helped by librarian Emma Milnes, who dug out some records from Victorian times that I found fascinating and helpful. I am also extremely grateful for the help and friendship shown by the Zoo's press officer Tina Campanella, who dug out some of the glorious photos in this book as well as occasionally letting me sneak into the zoo for a tryst with Marilyn. Tina also set up my valuable meeting with Dr Nisha Owen, who told me all about the work of the Zoological Society's EDGE programme towards ensuring the survival of the pygmy sloth on the island of Escudo de Varaguas, Panama.

The inclusion of such acknowledged sloth-lovers in the above is hardly surprising, but I should also mention two more serendipitous pieces of help that I received. First was my chance meeting with recording engineer Stuart Bruce, who happily listened to me chatting about sloths while he was fiddling with my microphones during recordings of Channel 4's *Gogglebox* programme. 'Would you like to see my sloth photographs?' he asked innocently, then showed me his private collection of pictures taken during a trip to Costa Rica in 2014. As you can see, I leapt on his generous offer to include several in these pages.

Even more surprising was an email I received from my Cambridge neighbour Robin Anderson just when I thought the book was finished. 'Thought you would be interested in this,' he wrote, adding a link to a report in the *Guardian*. And that was how I learnt the vital piece of information that completed my knowledge: sloths do not fart.

Finally, I must thank two other groups who have made writing this book so enjoyable. I am hugely grateful to all at Atlantic Books, especially managing director Will Atkinson, for appreciating how much the world needs an all-embracing book about sloths, my editor James Nightingale, who has done so much to help shape the book and make its writing such a pleasure, my copyeditor Nick Somogyi, whose corrections and suggestions were greatly appreciated, and all the others at Atlantic who sampled my sloth biscuits and made such complimentary comments.

Last, and perhaps most important of all, I must register my intense gratitude to my sons, James and Nick, without whom I might never have discovered the YouTube videos that led to my love of sloths and experienced the great joy this entire project has given me.

Illustration Credits

the author); Perfectly designed for hanging upside down (*seubsai/ Shutterstock.com*); Choloepas claws (*Inspired by Maps/Shutterstock. com*); Swimming sloth (*Martijn Smeets/Shutterstock.com*); Sloth on the road (*Scenic Shutterbug/Shutterstock.com*); Upwardly mobile (*Alina Lavrenova/Shutterstock.com*); Remarkable head rotation (*Arif Alakbar/Shutterstock.com*); Just hanging around (*Kristel Segeren/ Shutterstock.com*); Young sloth (*Kristel Segeren/Shutterstock.com*); Hoffman's two-toed sloth (*Ged. I MacKenzie/Shutterstock.com*)

Second colour section

Busy doing nothing (*Janossy Gergely/Shutterstock.com*); How could you call me a deadly sin? (*Kristel Segeren/Shutterstock.com*); The joy of hanging upside down in the wind (*Matthieu Gallet/Shutterstock. com*); In the wild in Costa Rica (*Kristel Segeren/Shutterstock. com*); A male three-toed sloth in Costa Rica (*Tanguy de Saint-Cyr/ Shutterstock.com*); On its way down… (*Jacek Sledzinski/Shutterstock. com*); Three legs hanging, one arm foraging (*Tami Freed/Shutterstock. com*); A sloths claws… (*Ryan.R.Smith.87/Shutterstock.com*); Three legs hanging, one arm eating (*Courtesy of Stuart Bruce*); Getting to know the ropes (*Courtesy of Stuart Bruce*); The wrong way to take a selfie (*Courtesy of Stuart Bruce*); A characteristic welcoming smile from a three-toed sloth (*Courtesy of Stuart Bruce*)

Third colour section

Marilyn with sloth baby at ZSL London Zoo's Rainforest Life exhibit, 2014 (© *ZSL London Zoo*); Baby Lento (© *ZSL London Zoo*); Warm and safe (© *ZSL London Zoo*); Mixed veg (© *ZSL London Zoo*);

Edward at ZSL London Zoo (© *ZSL London Zoo*); Hanging lesson (© *ZSL London Zoo*); Do not disturb (© *Nisha Owen ZSL*); We can be very fussy (© *Nisha Owen ZSL*); Nisha Owen of ZSL calming a pygmy sloth (© *Nisha Owen ZSL*); Fitting a radio collar (© *Nisha Owen ZSL*); 'I'll do my best...' (© *Nisha Owen ZSL*); Pygmy sloth proudly displaying its tracking collar (© *Nisha Owen ZSL*); 'How sweet to be a sloth' (© *Nisha Owen ZSL*); Sloth at sunset (*Prazis Images/Shutterstock.com*)

While every effort has been made to locate copyright-holders of illustrations, the author and publisher would be grateful for information about any illustrations where they have been unable to trace them, and would be glad to make amendments in further editions.

INDEX